普通高等学校人工智能交叉学科类规划教材

多机器人协同优化与 ROS 2 实践

Multi-Robot Cooperative Optimization
with ROS 2 Practice

刘庆山　赵　妍　编著

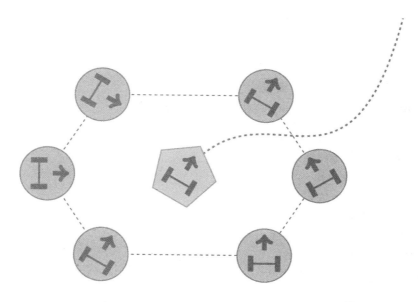

中国科学技术大学出版社

内 容 简 介

本书主要从优化的角度设计多机器人协同算法,并通过ROS 2实现Python 3程序代码.本书分别从理论分析、算法仿真和实物验证三个方面对非线性优化与多机器人协同进行理论介绍和实践演示.内容由浅入深,注重理论与实践相结合,既有理论方面的系统性介绍,又有结合ROS 2和Python 3代码的工程实践.期望能在多机器人协同优化的理论和实践研究中起到抛砖引玉的作用.

本书既可作为对非线性优化和多机器人编队感兴趣的研究人员的入门自学材料,又可作为高等院校相关专业的本科高年级学生和研究生的教材.

图书在版编目(CIP)数据

多机器人协同优化与ROS 2实践/刘庆山,赵研编著. —合肥:中国科学技术大学出版社,2023.6
ISBN 978-7-312-05684-0

Ⅰ. 多⋯ Ⅱ. ①刘⋯ ②赵⋯ Ⅲ. 机器人技术—研究 Ⅳ. TP242

中国国家版本馆CIP数据核字(2023)第094530号

多机器人协同优化与ROS 2实践
DUO JIQIREN XIETONG YOUHUA YU ROS 2 SHIJIAN

出版	中国科学技术大学出版社
	安徽省合肥市金寨路96号,230026
	http://press.ustc.edu.cn
	https://zgkxjsdxcbs.tmall.com
印刷	合肥市宏基印刷有限公司
发行	中国科学技术大学出版社
开本	787 mm×1092 mm 1/16
印张	14.75
字数	350千
版次	2023年6月第1版
印次	2023年6月第1次印刷
定价	56.00元

前　　言

近年来, 随着人工智能理论和技术的不断发展, 有关机器人技术的研究, 无论是在学术界还是工业界, 都引起了越来越多的关注. 尤其是近年来随着机器人操作系统 (ROS 和 ROS 2) 的不断发展和完善, 助推了理论知识应用于机器人的实践研究, 同时促进了机器人领域的理论和应用的发展. 当前我们正处于第三次人工智能热潮中, 有关机器人操作系统的介绍和应用已经出版了大量图书. 然而, 关于多机器人系统协同优化大多数还处在相关研究人员的学术成果层面, 目前还没有这方面的实践图书出版发行.

本书力求实践多机器人系统的协同控制, 以多机器人编队为切入点, 以集中式和分布式优化算法的设计为理论基础, 以 ROS 2 为实践手段, 旨在让读者在实践中理解和把握多机器人协同优化的理论和方法. 一方面, 本书强调理论方面的基本原理, 通过选择分布式优化理论研究方面的最新成果, 以较简单的算法设计为基础, 让读者了解这一方向研究的最新成果以帮助读者把握相关的理论方法. 另一方面, 本书更强调算法的具体实现, 通过大量的代码并对其进行解释和说明, 帮助读者快速将理论方法应用于实践. 建议读者在阅读本书时通过完成实际代码来学习.

本书对代码的实现包括仿真部分和实物部分, 读者在掌握了仿真程序后, 快速地将仿真程序进行简单的修改便可以应用于实物实现. 仿真的目的是快速掌握对算法理论的理解, 并快速整理算法的实现框架. 建议读者在掌握仿真算法的同时, 能够在实物上实现相应的算法.

本书对算法的实现是基于 Python 3 程序设计语言以及 ROS 2 Humble Hawksbill 版本, 所以在内容上是紧跟当前科研和行业发展前沿的; 同时还对 ROS 2 的基本知识做了详细解释, 所以也可作为对 ROS 2 感兴趣的读者的入门书.

由于本书包含一些优化理论方面的知识, 所以需要读者掌握一定的高等数学和线性代数的基本知识, 这样可以帮助读者快速理解和掌握本书中的理论部分. 此外, 本书使用 Python 3 作为编程语言来实现相关程序，即使对 Python 3 不熟悉的读者也能轻松地理解本书中的程序. 但是，由于篇幅限制，书中并没有对所有的代码细节都进行解释，如果你是 Python 初学者，建议先阅读 Python 的入门图书并掌握 Python 的基本语法结构.

本书分别从理论和实践方面对多机器人协同优化进行分析和代码实现. 内容主要包括 3 个部分: 第 1 部分是理论篇，包括 4 章内容: 第 1 章介绍凸优化的相关知识; 第 2 章介绍具有固定相对位置的多机器人编队问题; 第 3 章介绍基于几何相似性的多机器人编队问题; 第 4 章介绍时变分布式优化与多机器人实时围堵问题. 第 2 部分是仿真篇，包括 5 章内容: 第 5 章介绍 ROS 2 的一些基本知识，通过 Python 3 代码帮助读者快速熟悉 ROS 2 的基本架构; 第 6 章介绍 ROS 2 的坐标变换知识，这部分知识在多机器人编队实现中具有重要的意义; 第 7 章介绍基于集中式和分布式优化算法的 ROS 2 仿真; 第 8 章介绍基于几何相似性的多机器人编队仿真; 第 9 章介绍基于时变分布式优化的多机器人围堵仿真. 第 3 部分是实物篇，包括 1 章内容: 第 10 章介绍基于分布式优化的多机器人实物编队. 附录介绍了微分系统、李雅普诺夫稳定性理论和拉萨尔不变原理的一些基本概念和理论结果，以及机器人控制器的设计方法，读者通过阅读这部分知识可以更好地理解书中的理论部分.

如果读者需要本书代码的电子版本，可以发送邮件至 zhaoyan_503@126.com 索取.

编 者

2023 年 5 月

目　录

前言 ··· i

第 1 部分　理　论　篇

第 1 章　凸优化理论 ·· **2**
 1.1　凸集 ··· 2
 1.2　点向凸集的投影 ··· 3
  1.2.1　点向一般闭凸集的投影 ·· 3
  1.2.2　两个特殊投影算子 ·· 4
 1.3　凸函数 ·· 4
 1.4　凸优化问题 ··· 6
  1.4.1　无约束优化 ·· 6
  1.4.2　约束优化 ··· 7
  1.4.3　最优性条件 ·· 7
 1.5　变分不等式问题 ··· 8
 1.6　对偶理论与鞍点定理 ·· 9
 练习 ·· 10

第 2 章　具有固定相对位置的多机器人编队 ·· **11**
 2.1　编队问题 ·· 11
 2.2　具有固定相对位置的队形 ··· 11
  2.2.1　有参考中心的编队优化模型 ·· 13
  2.2.2　无参考中心的编队优化模型 ·· 15
 2.3　集中式优化算法 ··· 17
 2.4　分布式优化算法 ··· 22
 练习 ·· 23

第 3 章　基于几何相似性的多机器人编队 ··· **24**
 3.1　向量的几何变换 ··· 24
  3.1.1　旋转变换 ··· 24

	3.1.2 伸缩变换	24
	3.1.3 平移变换	24
	3.1.4 混合变换	25
3.2	实现多机器人编队的约束处理	25
	3.2.1 队形约束的简化处理	26
	3.2.2 带锚点的队形约束	28
3.3	多机器人编队的优化问题	29
3.4	集中式优化算法	29
3.5	分布式优化算法	30
练习		35

第 4 章　时变分布式优化与多机器人实时围堵　　36

4.1	多机器人实时围堵问题	36
4.2	集中式优化算法	40
4.3	分布式优化算法	42
练习		44

第 2 部分　仿　真　篇

第 5 章　ROS 2 简介　　46

5.1	版本选择	46
5.2	安装并测试 ROS 2	47
5.3	创建并初始化本书的工作目录	48
5.4	创建第一个功能包	49
5.5	消息	49
5.6	服务	56
5.7	动作	61
练习		67

第 6 章　ROS 2 坐标变换　　68

6.1	实现小海龟的领导-跟随任务	68
	6.1.1 小海龟状态发布	68
	6.1.2 跟随目标状态发布	72
	6.1.3 跟随小海龟控制	74
	6.1.4 使用 launch 文件启动多个节点	77
6.2	在 Gazebo 仿真中实现 TurtleBot 3 的领导-跟随任务	81
	6.2.1 TurtleBot 3 状态发布	81
	6.2.2 跟随目标状态发布	83
	6.2.3 跟随机器人控制	84

		6.2.4 使用 launch 文件启动多个节点	86
练习			92

第 7 章 基于集中式和分布式优化算法的 ROS 2 仿真 … 93

- 7.1 实现单个机器人的定点移动 … 93
  - 7.1.1 编写定点移动代码 … 93
  - 7.1.2 编写 launch 文件测试定点移动效果 … 96
- 7.2 多个机器人队形切换——集中式优化算法 … 100
  - 7.2.1 计算最优队形 … 101
  - 7.2.2 机器人根据最优队形定点移动 … 107
  - 7.2.3 编写 launch 文件测试队形切换效果 … 112
- 7.3 多个机器人队形切换——分布式优化算法 … 116
  - 7.3.1 分布式计算最优队形和机器人控制 … 118
  - 7.3.2 编写 launch 文件测试队形切换效果 … 128
- 练习 … 132

第 8 章 基于几何相似性的多机器人编队仿真 … 133

- 8.1 算法设计框架 … 133
- 8.2 代码实现 … 135
  - 8.2.1 节点代码实现 … 135
  - 8.2.2 编写 launch 文件测试编队切换效果 … 150
- 练习 … 156

第 9 章 基于时变分布式优化的多机器人围堵仿真 … 158

- 9.1 时变分布式优化与实时求解算法 … 158
- 9.2 分布式优化算法实现多机器人围堵任务 … 159
  - 9.2.1 实现机器人实时围堵控制 … 160
  - 9.2.2 编写 launch 文件测试实时围堵效果 … 166
- 练习 … 170

第 3 部分 实 物 篇

第 10 章 基于分布式优化的多机器人实物编队 … 172

- 10.1 轮式里程计 … 172
  - 10.1.1 利用轮式里程计实现机器人定位 … 174
  - 10.1.2 编写定点移动代码 … 177
  - 10.1.3 编写 launch 文件测试定点移动效果 … 180
- 10.2 ROS 2 导航工具包 Nav 2 … 183
  - 10.2.1 利用 Nav 2 编写定点移动代码 … 184
  - 10.2.2 编写 launch 文件测试定点移动效果 … 186

10.3　利用 Nav 2 实现多机器人队形切换 ·· 190
　　10.3.1　第一个机器人节点文件 ·· 191
　　10.3.2　第一个机器人 launch 文件 ·· 204
　　10.3.3　第一个机器人配置文件 ·· 208
　　10.3.4　实物测试 ·· 210
　练习 ·· 211

附录 A　微分系统 ·· **212**

附录 B　李雅普诺夫稳定性理论 ·· **216**

附录 C　拉萨尔不变原理 ·· **220**

附录 D　机器人控制器设计 ·· **223**

参考文献 ·· 225

索引 ·· 227

后记 ·· 228

第1部分

理 论 篇

第 1 章　凸优化理论

优化问题概括起来可以分为两类: 凸优化问题和非凸优化问题。在实际应用中, 我们遇到的很多问题都是非凸的. 从优化求解的角度来看, 非凸优化问题求解一般比较困难. 然而, 对于凸优化问题, 其理论基础则相对比较完善. 本章中, 将重点介绍凸优化的一些相关概念和结论.

1.1　凸　　集

在回答什么是凸优化问题之前, 先介绍什么是凸集 (convex set) 和凸函数 (convex function). 集合论是现代数学的一个基本的分支学科, 是构成数学公理化的基础之一, 其基本研究对象是一般集合. 把具有某种特定性质的对象的总体称为**集合**, 这个总体中的对象称为该集合的**元素**. 集合论在数学中占有一个独特的地位, 它的基本概念已经渗透到数学的所有领域. 我们说凸集是一类特殊的集合, 其定义如下所述:

定义 1.1 (凸集)

欧几里得空间 $\mathbf{R}^n$ 的子集 Ω 称为凸集, 如果其满足

$$\alpha \boldsymbol{x} + (1-\alpha)\boldsymbol{y} \in \Omega, \quad \forall \boldsymbol{x}, \boldsymbol{y} \in \Omega, \forall \alpha \in [0,1] \tag{1.1}$$

注　从几何上看, 一个集合称为**凸集**, 是指这个集合中的任意两个点的连线仍然在这个集合中.

图 1.1 给出了平面上凸集与非凸集的例子: 左图是凸集, 右图是非凸集. 此外, 超平面 $H = \{\boldsymbol{x} \in \mathbf{R}^n | \boldsymbol{p}^{\mathrm{T}} \boldsymbol{x} = q\}$ 也是凸集, 其中非零向量 $\boldsymbol{p}$ 称为超平面的法向量, q 是实数. 集合 $S = \{\boldsymbol{x} \in \mathbf{R}^n | \boldsymbol{A}\boldsymbol{x} = \boldsymbol{b}\}$ 是 n 维欧几里得空间中多个超平面的交集, 也是凸集, 其中矩阵 $\boldsymbol{A} \in \mathbf{R}^{m \times n}$ $(m < n)$, $\boldsymbol{b} \in \mathbf{R}^m$.

注　如果一个凸集同时也是闭集, 我们称这个凸集为**闭凸集**.

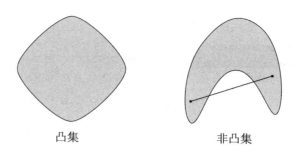

图 1.1 凸集和非凸集的例子

1.2 点向凸集的投影

接下来, 本节研究欧几里得空间中任意一个点向一个闭凸集的投影.

1.2.1 点向一般闭凸集的投影

给出点向一般闭凸集的投影定义如下:

定义 1.2 (投影算子)

设 Ω 是欧几里得空间 $\mathbf{R}^n$ 的一个闭凸集, 我们定义从 $\mathbf{R}^n$ 到 Ω 的投影算子为如下映射:
$$g(\boldsymbol{u}) = \arg\min_{\boldsymbol{v} \in \Omega} \|\boldsymbol{u} - \boldsymbol{v}\| \tag{1.2}$$
其中 $\arg\min\limits_{\boldsymbol{v} \in \Omega} \|\cdot\|$ 表示当范数 (距离) 达到最小值时的点的坐标.

注 从几何上看, $g(\boldsymbol{u})$ 即是 Ω 中到点 $\boldsymbol{u}$ 距离最近的点. 当 $\boldsymbol{u} \in \Omega$ 时, $g(\boldsymbol{u}) = \boldsymbol{u}$; 当 $\boldsymbol{u} \notin \Omega$ 时, $g(\boldsymbol{u})$ 是 Ω 边界上的距离 $\boldsymbol{u}$ 最近的某个点, 如图1.2所示.

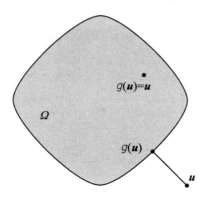

图 1.2 平面上点 $\boldsymbol{u}$ 向闭凸集 Ω 的投影

关于投影算子, 有如下常用的结论:

> **定理 1.1 (投影定理 (文献 [16](定理 1.2.5)))**
> 设 Ω 是 $\mathbf{R}^n$ 上的非空闭凸集, 则有如下结论成立:
> (1) 公式(1.2)中的 $g(\boldsymbol{u})$ 是唯一的, 即存在唯一的 $\bar{\boldsymbol{u}} = g(\boldsymbol{u})$, 使得
> $$\|\bar{\boldsymbol{u}} - \boldsymbol{u}\| = \min_{\boldsymbol{v} \in \Omega} \|\boldsymbol{u} - \boldsymbol{v}\|$$
> (2) $\bar{\boldsymbol{u}} \in \Omega$ 满足 $\bar{\boldsymbol{u}} = g(\boldsymbol{u})$ 的充分必要条件是
> $$(\boldsymbol{u} - \bar{\boldsymbol{u}})^{\mathrm{T}}(\bar{\boldsymbol{u}} - \boldsymbol{v}) \geqslant 0, \quad \forall \boldsymbol{v} \in \Omega$$

通常情况下, 对于任意一个闭凸集, 其上的投影算子很难写出其解析表达式. 然而, 对于某些特殊的闭凸集, 可以写出其投影算子的解析表达式.

1.2.2 两个特殊投影算子

下面考虑两个特殊的闭凸集, 一个是盒形闭凸集 $\Omega_1 = \{\boldsymbol{x} \in \mathbf{R}^n | l_i \leqslant x_i \leqslant h_i (i = 1, 2, \cdots, n)\}$, 另一个是球形闭凸集 $\Omega_2 = \{\boldsymbol{x} \in \mathbf{R}^n | \|\boldsymbol{x} - \boldsymbol{s}\| \leqslant r\}$, 其中 $\boldsymbol{s} \in \mathbf{R}^n$ 是球心坐标, l_i, h_i, r 是给定的实数.

对于盒形闭凸集, 如果设定义1.2中的 $g(\boldsymbol{u}) = (g_1(u_1), g_2(u_2), \cdots, g_n(u_n))^{\mathrm{T}}$, 则每个 $g_i(u_i)$ 是一个分段线性函数 (图1.3的左图), 其表达式为

$$g_i(u_i) = \begin{cases} l_i, & u_i < l_i \\ u_i, & l_i \leqslant u_i \leqslant h_i \\ h_i, & u_i > h_i \end{cases} \tag{1.3}$$

同样地, 对于球形闭凸集 (图1.3的右图), $g(\boldsymbol{u})$ 的表达式为

$$g(\boldsymbol{u}) = \begin{cases} \boldsymbol{u}, & \|\boldsymbol{u} - \boldsymbol{s}\| \leqslant r \\ \boldsymbol{s} + r\dfrac{\boldsymbol{u} - \boldsymbol{s}}{\|\boldsymbol{u} - \boldsymbol{s}\|}, & \|\boldsymbol{u} - \boldsymbol{s}\| > r \end{cases} \tag{1.4}$$

1.3 凸 函 数

在介绍了凸集的基本概念之后, 下面来看看凸函数的概念.

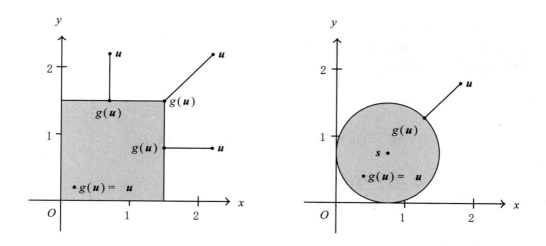

图 1.3　盒形和球形闭凸集上的投影算子示意图

> **定义 1.3 (凸函数)**
>
> 如果函数 f 的定义域 D 是凸集,且有
>
> $$f(\alpha \boldsymbol{x}+(1-\alpha)\boldsymbol{y}) \leqslant \alpha f(\boldsymbol{x})+(1-\alpha)f(\boldsymbol{y}), \quad \forall \boldsymbol{x},\boldsymbol{y}\in D, \forall \alpha\in[0,1] \qquad (1.5)$$
>
> 则函数 $f:\mathbf{R}^n \to \mathbf{R}$ 称为凸函数. 如果上式对 $\forall \boldsymbol{x}\neq \boldsymbol{y}$ 和 $\forall \alpha\in(0,1)$ 严格不等式成立,则称 f 是凸集 D 上的严格凸函数.

注　从几何上看,一个函数是**凸函数**,是指连接函数 f 图像上任意两个点 $A(\boldsymbol{x},f(\boldsymbol{x}))$ 和 $B(\boldsymbol{y},f(\boldsymbol{y}))$ 的线段位于函数图像 (曲线或曲面) 的上方 (图1.4).

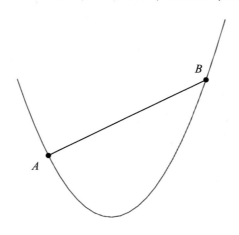

图 1.4　凸函数示意图

连接 A 点和 B 点的线段在函数图像 (具有相同的自变量区间) 的上方.

如果 $f(\boldsymbol{x})$ 在凸集 D 上有连续一阶偏导数，则 $f(\boldsymbol{x})$ 具有如下的性质：

定理 1.2 (文献 [10] (3.1.3 节))
函数 $f: \mathbf{R}^n \to \mathbf{R}$ 在凸集 D 上有连续一阶偏导数，则
(1) $f(\boldsymbol{x})$ 是 D 上凸函数的充分必要条件是
$$f(\boldsymbol{y}) - f(\boldsymbol{x}) \geqslant (\boldsymbol{y}-\boldsymbol{x})^\mathrm{T} \nabla f(\boldsymbol{x}), \quad \forall \boldsymbol{x},\boldsymbol{y} \in D$$
(2) $f(\boldsymbol{x})$ 是 D 上严格凸函数的充分必要条件是
$$f(\boldsymbol{y}) - f(\boldsymbol{x}) > (\boldsymbol{y}-\boldsymbol{x})^\mathrm{T} \nabla f(\boldsymbol{x}), \quad \forall \boldsymbol{x},\boldsymbol{y} \in D, \boldsymbol{x} \neq \boldsymbol{y}$$
其中 ∇f 表示函数 f 的梯度.

1.4 凸优化问题

有了前面的凸集和凸函数的概念后，接下来介绍凸优化的相关概念和性质.

1.4.1 无约束优化

在凸优化问题中，形式上最简单的一类凸优化问题是无约束凸优化问题，其一般形式为

$$\min \quad f(\boldsymbol{x}) \tag{1.6}$$

其中 $f: \mathbf{R}^n \to \mathbf{R}$ 是凸函数. 若 $f(\boldsymbol{x})$ 在 $\mathbf{R}^n$ 上是连续可微的，则 $\boldsymbol{x}^*$ 是上述无约束优化问题的最优解的充要条件是 $\nabla f(\boldsymbol{x}^*) = 0$. 求解问题 (1.6) 的一个常用的方法是**梯度下降法**，或者称为**最速下降法**，其一般算法流程如下：

算法 1.1 (梯度下降法)
步骤 1: 给定初始点 $\boldsymbol{x}(0) \in \mathbf{R}^n$ 和迭代误差 $\epsilon > 0$；
步骤 2: 计算目标函数在 $\boldsymbol{x}(k)$ 点的梯度 $\nabla f(\boldsymbol{x}(k))$，如果 $\|\nabla f(\boldsymbol{x}(k))\| < \epsilon$，则算法结束，否则转步骤 3；
步骤 3: 计算 $\boldsymbol{x}(k+1) = \boldsymbol{x}(k) - \sigma_k \nabla f(\boldsymbol{x}(k))$；
步骤 4: $k := k+1$，转步骤 2.

注 算法 1.1 中的 σ_k 是迭代步长，可以通过精确直线搜索或者非精确直线搜索的方法确定 [10].

1.4.2 约束优化

在凸优化方法的实际工程应用中,约束条件是普遍存在的.因此,需要研究更一般的约束凸优化问题.考虑如下的约束优化问题:

$$\min \quad f(\boldsymbol{x}) \\ \text{s.t.} \begin{cases} \boldsymbol{A}\boldsymbol{x} = \boldsymbol{b} \\ \boldsymbol{c}(\boldsymbol{x}) \leqslant 0 \end{cases} \tag{1.7}$$

其中目标函数 $f : \mathbf{R}^n \to \mathbf{R}$ 是凸函数;$\boldsymbol{A} \in \mathbf{R}^{m \times n}$ 是行满秩矩阵,$\boldsymbol{b} \in \mathbf{R}^m$;$\boldsymbol{c}(\boldsymbol{x}) = (c_1(x), c_2(x), \cdots, c_s(x))^{\mathrm{T}}$,$c_i : \mathbf{R}^n \to \mathbf{R}$ 是不等式约束函数,也是凸函数.我们称如下集合:

$$X = \{\boldsymbol{x} \in \mathbf{R}^n | \boldsymbol{A}\boldsymbol{x} = \boldsymbol{b}, \boldsymbol{c}(\boldsymbol{x}) \leqslant 0\}$$

为优化问题(1.7)的**可行域**(feasible region).可行域即是满足所有约束条件的点构成的集合.对于凸优化问题来说,其可行域是凸集.

对于约束优化问题,由可行域的定义可知,求解约束优化问题即是在可行域上寻找一个点 $\boldsymbol{x}^*$,使得目标函数 $f(\boldsymbol{x}^*)$ 达到最小.因此,利用可行域的概念,可以将问题(1.7)写成如下的等价形式:

$$\min \quad f(\boldsymbol{x}) \\ \text{s.t.} \quad \boldsymbol{x} \in X \tag{1.8}$$

对于约束优化问题(1.7)的理论研究,通常要求其可行域满足如下的 **Slater 条件**:

> **定义 1.4 (Slater 条件)**
> 存在一点 $\hat{\boldsymbol{x}} \in X$ 使得如下等式和不等式成立:
> $$\boldsymbol{A}\hat{\boldsymbol{x}} = \boldsymbol{b}$$
> $$\boldsymbol{c}(\hat{\boldsymbol{x}}) < 0$$

1.4.3 最优性条件

接下来,讨论约束优化问题的一阶最优性条件,它通常被称为 KKT(Karush-Kuhn-Tucker) 条件.

> **定理 1.3 (KKT 条件 (文献 [10](5.5.3 节)))**
>
> 设 $\boldsymbol{x}^*$ 是凸优化问题(1.7)的可行解,且 $f(\boldsymbol{x})$ 和 $c_j(\boldsymbol{x})$ 在 $\boldsymbol{x}^*$ 的领域内一阶连续可微,则 $\boldsymbol{x}^*$ 是最优解的充要条件是存在 λ_i 和 $\mu_j \geqslant 0$ 使得
>
> $$\nabla f(\boldsymbol{x}^*) + \sum_{i=1}^m \lambda_i \boldsymbol{a}_i^{\mathrm{T}} + \sum_{j=1}^s \mu_j \nabla c_j(\boldsymbol{x}^*) = 0$$
> $$\mu_j c_j(\boldsymbol{x}^*) = 0, \quad j = 1, 2, \cdots, s,$$
>
> 其中 $\boldsymbol{a}_i$ 是矩阵 $\boldsymbol{A}$ 的第 i 行.

1.5 变分不等式问题

与优化问题密切相关的一类问题是变分不等式问题. 此类问题描述如下: 设 Ω 是 $\mathbf{R}^n$ 中的闭凸集, F 是 $\Omega \to \mathbf{R}^n$ 的映射, **变分不等式**问题即是寻找 $\boldsymbol{x}^* \in \Omega$ 使得如下的不等式成立:

$$(\boldsymbol{x} - \boldsymbol{x}^*)^{\mathrm{T}} F(\boldsymbol{x}^*) \geqslant 0, \quad \forall \boldsymbol{x} \in \Omega \tag{1.9}$$

> **定义 1.5 (单调映射)**
>
> 设 Ω 是 $\mathbf{R}^n$ 中的闭凸集, 如果映射 $F: \Omega \to \mathbf{R}^n$ 满足
>
> $$(\boldsymbol{x} - \boldsymbol{y})^{\mathrm{T}} (F(\boldsymbol{x}) - F(\boldsymbol{y})) \geqslant 0, \quad \forall \boldsymbol{x}, \boldsymbol{y} \in \Omega$$
>
> 则称其为单调映射. 进而, 如果上述等式仅在 $\boldsymbol{x} = \boldsymbol{y}$ 时成立, 我们称 F 为严格单调映射.

如果问题(1.9)中的映射 F 是单调映射, 则称其为**单调变分不等式**问题. 一般情况下, 如果函数 $f(\boldsymbol{x})$ 是凸函数, 则其梯度 $\nabla f(\boldsymbol{x})$ 是单调的. 关于凸优化问题(1.8)和变分不等式问题(1.9), 有如下的重要结论:

> **定理 1.4 (文献 [2](Chapter 1, Propositions 5.1~5.2))**
>
> 函数 $f: \mathbf{R}^n \to \mathbf{R}$ 在可行域 X 上有连续一阶偏导数, $\boldsymbol{x}^*$ 是优化问题(1.8)的解, 当且仅当 $\boldsymbol{x}^*$ 是如下变分不等式问题的解:
>
> $$(\boldsymbol{x} - \boldsymbol{x}^*)^{\mathrm{T}} \nabla f(\boldsymbol{x}^*) \geqslant 0, \quad \forall \boldsymbol{x} \in X$$

接下来, 对于变分不等式问题(1.9), 有下列结论成立:

定理 1.5 (文献 [7] (Proposition 3.1))

$x^* \in \Omega$ 是变分不等式问题(1.9)的解当且仅当 x^* 满足如下的投影方程:

$$x^* = g(x^* - \sigma F(x^*))$$

其中 $\sigma > 0$ 是任意实数, $g(\cdot)$ 是定义1.2中定义的投影算子.

1.6 对偶理论与鞍点定理

对偶理论为求解约束优化问题提供了重要的理论支持, 这一节介绍这一理论以及与之相关的鞍点定理.

定义 1.6 (鞍点)

点 (x^*, y^*) 是函数 $f(x, y): X \times Y \to \mathbf{R}$ 的一个鞍点(关于 X 极小化和关于 Y 极大化), 是指对于任意的 $(x, y) \in X \times Y$, 有

$$f(x^*, y) \leqslant f(x^*, y^*) \leqslant f(x, y^*)$$

考虑如下的约束优化问题:

$$\min \quad f(x)$$
$$\text{s.t.} \begin{cases} Ax = b \\ c(x) \leqslant 0 \\ x \in \Omega \end{cases} \tag{1.10}$$

其中 Ω 是闭凸集, 其他符号同(1.7)中的定义. 问题(1.10)的**拉格朗日函数**(Lagrangian function) 定义为

$$L(x, \lambda, \mu) = f(x) + \lambda^{\mathrm{T}}(Ax - b) + \mu^{\mathrm{T}}c(x) \tag{1.11}$$

其中 $\lambda \in \mathbf{R}^m, \mu \in \mathbf{R}^s$ 且 $\mu \geqslant 0$.

因为

$$\sup_{\mu \geqslant 0} L(x, \lambda, \mu) = \begin{cases} f(x), & Ax - b = 0, c(x) \leqslant 0 \\ +\infty, & \text{其他} \end{cases}$$

问题(1.10)可以表示为

$$\min_{x \in \Omega} \max_{\mu \geqslant 0} L(x, \lambda, \mu)$$

定义函数

$$d(\boldsymbol{\lambda}, \boldsymbol{\mu}) = \min_{\boldsymbol{x} \in \Omega} L(\boldsymbol{x}, \boldsymbol{\lambda}, \boldsymbol{\mu})$$

则(1.10)的**对偶问题**(dual problem) 可以写成下列优化问题:

$$\max_{\boldsymbol{\mu} \geqslant 0} d(\boldsymbol{\lambda}, \boldsymbol{\mu}) = \max_{\boldsymbol{\mu} \geqslant 0} \min_{\boldsymbol{x} \in \Omega} L(\boldsymbol{x}, \boldsymbol{\lambda}, \boldsymbol{\mu}) \tag{1.12}$$

关于鞍点与约束优化问题的关系, 有如下**鞍点定理**成立:

> **定理 1.6 (鞍点定理 (文献 [18] (定理 7.3.4)))**
>
> 设 $(\boldsymbol{x}^*, \boldsymbol{\lambda}^*, \boldsymbol{\mu}^*)$ 是优化问题(1.10)的拉格朗日函数(1.11)的鞍点, 当且仅当 $\boldsymbol{x}^*$ 和 $(\boldsymbol{\lambda}^*, \boldsymbol{\mu}^*)$ 分别是优化问题(1.10)和对偶问题(1.12)的最优解.

练 习

1. 用梯度下降法求解下列无约束优化问题:

$$\min \quad f(x_1, x_2) = x_1^2 + x_2^2 + x_1 x_2 - x_2$$

初始点取 $\boldsymbol{x}(0) = (0,0)^{\mathrm{T}}$, 完成如下两项任务:
  (1) 手动计算, 迭代两次;
  (2) 编写 Python 程序, 计算误差满足 $\|\nabla f\| < 10^{-3}$.

2. 用 KKT 条件求解下列非线性规划问题:

$$\min \quad 2x_1 + 3x_2 + 4x_3$$
$$\text{s.t.} \begin{cases} x_1^2 - x_2 - x_3 \leqslant 0 \\ x_1 - 1 \geqslant 0 \\ x_3 + 1 \geqslant 0 \end{cases}$$

3. 写出如下线性规划问题的拉格朗日对偶问题:

$$\min \quad \boldsymbol{c}^{\mathrm{T}} \boldsymbol{x}$$
$$\text{s.t.} \begin{cases} \boldsymbol{A}\boldsymbol{x} = \boldsymbol{b} \\ \boldsymbol{x} \geqslant 0 \end{cases}$$

其中 $\boldsymbol{x} \in \mathbf{R}^n, \boldsymbol{c} \in \mathbf{R}^n, \boldsymbol{A} \in \mathbf{R}^{m \times n}, \boldsymbol{b} \in \mathbf{R}^m$.

第 2 章 具有固定相对位置的多机器人编队

本章将介绍一类较为简单的多机器人编队问题, 即机器人之间具有固定相对位置的编队问题. 首先将建立实现这类编队问题的优化模型, 进而设计解决此类问题的优化算法. 我们主要考虑两种情况下的固定队形: 一种是有参考中心的队形, 另一种是无参考中心的队形. 并根据这两种队形设计相应的集中式和分布式优化算法.

2.1 编队问题

多机器人的编队问题主要包括两种: 一种是队形保持, 另一种是队形切换. 所谓**队形保持**, 是指机器人群在运动过程中需要保持一个固定的队形, 这个队形在机器人群运动过程中保持不变. 所谓**队形切换**, 是指机器人群在运动过程中需要进行队形的改变, 并且需要遵循一定的原则, 如队形切换时间最短, 整体能量消耗最少, 整体移动距离最短, 等等. 然而, 队形保持和队形切换的概念并不是完全不同的. 如果机器人群在运动过程中的队形是实时的, 那么这类队形保持问题也可以认为是队形切换问题.

本书中所说的队形 (formation),是指具有如下的形象化定义:

定义 2.1 (队形)
m 个机器人组成的队形, 是指由 m 个两两不重合点构成的具有一定几何形状的图标.

图2.1给出了两个队形的例子.

2.2 具有固定相对位置的队形

接下来, 考虑机器人两两之间具有固定相对位置的编队问题. 首先, 给出如下的一个有用的引理:

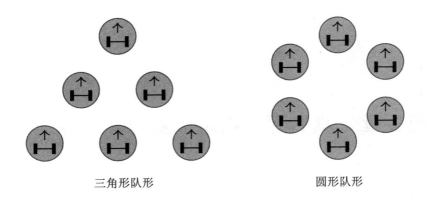

三角形队形　　　　　圆形队形

图 2.1　六辆无人车形成的两个队形的例子

> **引理 2.1**
> 设 $L \in \mathbf{R}^{mm}$ 是 m 个节点构成的无向连通图 $\mathcal{G}$ 的拉普拉斯矩阵,且 $\tilde{L} = L \otimes I_n \in \mathbf{R}^{mn \times mn}$,其中 I_n 是 n 维单位矩阵,$\otimes$ 是克罗内克 (Kronecker) 积. 那么 $\zeta_1 = \zeta_2 = \cdots = \zeta_m$ 当且仅当 $\tilde{L}\tilde{\zeta} = 0$,其中 $\zeta_i \in \mathbf{R}^n$ 是连通图中节点 i 的状态且满足 $\tilde{\zeta} = (\zeta_1^{\mathrm{T}}, \zeta_2^{\mathrm{T}}, \cdots, \zeta_m^{\mathrm{T}})^{\mathrm{T}}$.

证明　(必要性) 因为图 $\mathcal{G}$ 是连通的,其拉普拉斯矩阵 L 有单重零特征值,且对应的特征向量是 $\mathbf{1}_m$(元素全 1 的 m 维列向量). 设 $\zeta \in \mathbf{R}^n$ 是任意非零向量. 如果 $\tilde{\zeta} = \mathbf{1}_m \otimes \zeta$,有

$$\tilde{L}\tilde{\zeta} = (L \otimes I_n)\tilde{\zeta} = (L \otimes I_n)(\mathbf{1}_m \otimes \zeta)$$
$$= (L\mathbf{1}_m) \otimes (I_n\zeta) = 0$$

(充分性) 根据克罗内克积的性质:$\mathrm{vec}(AZB) = (B^{\mathrm{T}} \otimes A)\mathrm{vec}(Z)$,其中 A, B 和 Z 是具有相应维数的矩阵,$\mathrm{vec}(\cdot)$ 是将矩阵按列进行向量化. 因此 $\tilde{L}\tilde{\zeta} = (L \otimes I_n)\tilde{\zeta} = (L \otimes I_n)\mathrm{vec}(\zeta) = \mathrm{vec}(I_n\zeta L)$,其中 $\zeta = (\zeta_1, \zeta_2, \cdots, \zeta_m) \in \mathbf{R}^{n \times m}$. 如果 $\tilde{L}\tilde{\zeta} = 0$,有 $I_n\zeta L = 0$,进而可以推出 $L\zeta^{\mathrm{T}} = 0$. 因为图 $\mathcal{G}$ 是连通的,$\mathbf{1}_m$ 是矩阵 L 相应于单重特征值零的特征向量. 进而 $\zeta^{\mathrm{T}} = \alpha^{\mathrm{T}} \otimes \mathbf{1}_m$,其中 $\alpha = (\alpha_1, \alpha_2, \cdots, \alpha_n)^{\mathrm{T}} \in \mathbf{R}^n$,且 α_i $(i = 1, 2, \cdots, n)$ 是任意实数. 因此,我们有

$$\zeta = \alpha \otimes \mathbf{1}_m^{\mathrm{T}} = \begin{pmatrix} \alpha_1 \mathbf{1}_m^{\mathrm{T}} \\ \alpha_2 \mathbf{1}_m^{\mathrm{T}} \\ \vdots \\ \alpha_n \mathbf{1}_m^{\mathrm{T}} \end{pmatrix}$$

进一步可以得出,ζ 的每个列向量 ζ_i $(i = 1, 2, \cdots, m)$ 满足 $\zeta_i = \alpha$,因此 $\tilde{\zeta} = \mathrm{vec}(\zeta) = \mathbf{1}_m \otimes \alpha$.

综合上述分析,对于无向连通图 $\mathcal{G}$, $\tilde{L}\zeta = 0$ 当且仅当存在某个非零向量 $\alpha \in \mathbf{R}^n$ 使得 $\zeta_i = \alpha$ $(i = 1, 2, \cdots, m)$.   证毕!

2.2.1 有参考中心的编队优化模型

首先考虑有参考中心的多机器人编队问题,而这类问题可以分为两种情况:一种是队形中有现实参考中心,即队形中的某个机器人担任了参考中心的作用;另一种是队形中有虚拟参考中心,即队形中的所有机器人都不担任参考中心,而是有一个假设的虚拟参考中心. 图2.2给出了这两种情况的例子. 左图中标有 RC 的机器人是现实参考中心,右图中标有 VC 的点是虚拟参考中心.

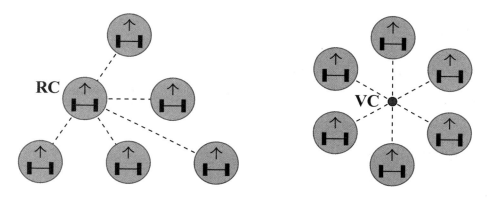

图 2.2   六辆无人车形成的有参考中心的两个队形的例子

1. 有现实参考中心的编队优化模型

假设有 m 个机器人需要完成编队任务,其中机器人 1 号担任参考中心的角色. 所有其他机器人与机器人 1 号的相对位置记为 Δ_{i1} $(i = 1, 2, \cdots, m)$,其中 $\Delta_{11} = 0$. 因此,如图2.2中左图所示,机器人队形需要满足约束条件

$$p_1 - \Delta_{11} = p_2 - \Delta_{21} = \cdots = p_i - \Delta_{i1} = \cdots = p_m - \Delta_{m1} \tag{2.1}$$

根据引理2.1,约束条件(2.1)等价于 $\tilde{L}(p - \Delta_1) = 0$,其中 $p = (p_1^{\mathrm{T}}, p_2^{\mathrm{T}}, \cdots, p_m^{\mathrm{T}})^{\mathrm{T}}$, $\Delta_1 = (\Delta_{11}^{\mathrm{T}}, \Delta_{21}^{\mathrm{T}}, \cdots, \Delta_{m1}^{\mathrm{T}})^{\mathrm{T}}$.

假设 m 个机器人的初始位置记为 $\xi_1, \xi_2, \cdots, \xi_m$ 是已知量;形成队形的目标位置记为 $p_1, p_2, \cdots, p_m$,是我们需要确定的未知量. 对于机器人 i,其初始位置到队形目标位置的移动距离为 $\|p_i - \xi_i\|$. 对于 m 个机器人,总的移动距离为 $\sum_{i=1}^{m} \|p_i - \xi_i\|$. 为了分析和算法实现上的方便,我们采用范数平方作为目标函数,其形式如下:

$$\frac{1}{2} \sum_{i=1}^{m} \|p_i - \xi_i\|^2$$

此目标函数具有连续光滑的梯度,系数 1/2 是为了保证求和中每一项 $\|\boldsymbol{p}_i - \boldsymbol{\xi}_i\|^2/2$ 关于 $\boldsymbol{p}_i$ 的梯度为 $\boldsymbol{p}_i - \boldsymbol{\xi}_i$,以方便理论分析.

此外,为了限制机器人在一个有限的空间范围内形成队形,我们给每个机器人添加一个运动范围的约束条件 $\boldsymbol{p}_i \in \Omega_i$,其中 Ω_i 是机器人 i 在二维或者三维空间中可移动的范围. 例如,若希望所有的机器人都在 $[0, 10] \times [0, 10]$ 的平面范围内形成队形,就可以设 $\Omega_i = [0, 10]^2$.

综合上述关于多机器人编队的目标函数和约束条件,得出如下的优化问题:

$$\min_{\boldsymbol{p}} \quad \frac{1}{2}\|\boldsymbol{p} - \boldsymbol{\xi}\|^2 \\ \text{s.t.} \begin{cases} \tilde{\boldsymbol{L}}(\boldsymbol{p} - \boldsymbol{\Delta}_1) = 0 \\ \boldsymbol{p} \in \Omega \end{cases} \tag{2.2}$$

其中 $\boldsymbol{p} = (\boldsymbol{p}_1^\mathrm{T}, \boldsymbol{p}_2^\mathrm{T}, \cdots, \boldsymbol{p}_m^\mathrm{T})^\mathrm{T}$ 表示 m 个机器人期望到达的目标队形的位置; $\boldsymbol{\xi} = (\boldsymbol{\xi}_1^\mathrm{T}, \boldsymbol{\xi}_2^\mathrm{T}, \cdots, \boldsymbol{\xi}_m^\mathrm{T})^\mathrm{T}$ 表示 m 个机器人的初始位置; $\Omega = \prod_{i=1}^{m} \Omega_i$ 表示 Ω_i 的笛卡儿乘积,用以限制机器人的移动范围.

2. 有虚拟参考中心的编队优化模型

上一节我们假设多机器人队形中有一个现实参考中心,然而,在实际应用中,如果现实参考中心发生故障,如遭受攻击、通信受阻等,有可能造成对整个编队的破坏. 下面考虑队形中有虚拟参考中心的情况.

同样地,假设有 m 个机器人需要完成编队任务,虚拟参考中心的坐标为 $\boldsymbol{p}_0$. 所有机器人与虚拟参考中心的相对位置记为 $\boldsymbol{\Delta}_{i0}$ ($i = 1, 2, \cdots, m$),是一个从虚拟参考中心指向机器人 i 的向量. 因此,如图2.2中右图所示,机器人队形需要满足如下约束条件:

$$\boldsymbol{p}_1 - \boldsymbol{\Delta}_{10} = \boldsymbol{p}_2 - \boldsymbol{\Delta}_{20} = \cdots = \boldsymbol{p}_i - \boldsymbol{\Delta}_{i0} = \cdots = \boldsymbol{p}_m - \boldsymbol{\Delta}_{m0} = \boldsymbol{p}_0 \tag{2.3}$$

根据引理2.1,约束条件(2.3)等价于 $\tilde{\boldsymbol{L}}(\boldsymbol{p} - \boldsymbol{\Delta}_0) = 0$,其中 $\boldsymbol{p} = (\boldsymbol{p}_1^\mathrm{T}, \boldsymbol{p}_2^\mathrm{T}, \cdots, \boldsymbol{p}_m^\mathrm{T})^\mathrm{T}$, $\boldsymbol{\Delta}_0 = (\boldsymbol{\Delta}_{10}^\mathrm{T}, \boldsymbol{\Delta}_{20}^\mathrm{T}, \cdots, \boldsymbol{\Delta}_{m0}^\mathrm{T})^\mathrm{T}$.

和上一节类似,对于有虚拟参考中心的多机器人编队,可以建立如下的优化问题:

$$\min_{\boldsymbol{p}} \quad \frac{1}{2}\|\boldsymbol{p} - \boldsymbol{\xi}\|^2 \\ \text{s.t.} \begin{cases} \tilde{\boldsymbol{L}}(\boldsymbol{p} - \boldsymbol{\Delta}_0) = 0 \\ \boldsymbol{p} \in \Omega \end{cases} \tag{2.4}$$

2.2.2 无参考中心的编队优化模型

对于有参考中心的多机器人编队问题,不管是现实参考中心,还是虚拟参考中心,如图2.2所示,其队形拓扑结构和通信拓扑结构可以是相互独立的. 即这类编队问题的实现,其物理层和通信层是完全不相关的. 那么,是否可以实现机器人队形的拓扑结构和通信结构的统一呢?接下来,将研究这类问题.

同样地,假设有 m 个机器人需要完成编队任务,其中没有中心节点作为现实或者虚拟参考中心. 然而,机器人之间通过两两互联形成一个连通网络,并且每个机器人通过通信可以知道自己的邻居的某些状态信息,如位置和速度等. 设机器人 i 和机器人 j 的相对位置记为 $\boldsymbol{\Delta}_{ij}$ $(i,j=1,2,\cdots,m)$. 因此,如图2.3所示,所有具有通信关系的机器人对需要满足如下的约束条件:

$$\boldsymbol{p}_i - \boldsymbol{p}_j = \boldsymbol{\Delta}_{ij} \tag{2.5}$$

其中具有通信关系的机器人知道自己和邻居的相对位置,所有机器人的相互连接网络形成一个连通的无向图.

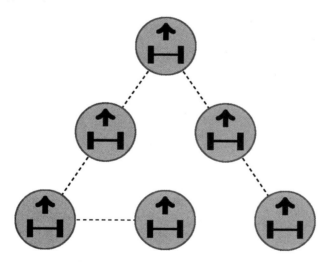

图 2.3 六个机器人形成三角形队形,具有通信关系的机器人知道自己和邻居的相对位置

下面将推导约束条件(2.5)等价于问题(2.4)中的等式约束. 事实上,可以假设队形中有一个参考中心,其对应的坐标为 $\boldsymbol{p}_0$. 由于机器人群构成了一个连通图,如果机器人 i 和 j 之间有通信,那么可以得出 $\boldsymbol{\Delta}_{ij} = \boldsymbol{p}_i - \boldsymbol{p}_j = \boldsymbol{p}_i - \boldsymbol{p}_0 - (\boldsymbol{p}_j - \boldsymbol{p}_0)$. 记 $\boldsymbol{\Delta}_{i0} = \boldsymbol{p}_i - \boldsymbol{p}_0$,则等式(2.5)等价于

$$\boldsymbol{p}_i - \boldsymbol{p}_j = \boldsymbol{\Delta}_{i0} - \boldsymbol{\Delta}_{j0}$$

即

$$\boldsymbol{p}_i - \boldsymbol{\Delta}_{i0} = \boldsymbol{p}_j - \boldsymbol{\Delta}_{j0} \tag{2.6}$$

根据引理2.1,约束条件(2.6)等价于 $\tilde{\boldsymbol{L}}(\boldsymbol{p} - \boldsymbol{\Delta}_0) = 0$,其中 $\boldsymbol{p} = (\boldsymbol{p}_1^{\mathrm{T}}, \boldsymbol{p}_2^{\mathrm{T}}, \cdots, \boldsymbol{p}_m^{\mathrm{T}})^{\mathrm{T}}$,

$\boldsymbol{\Delta}_0 = (\boldsymbol{\Delta}_{10}^{\mathrm{T}}, \boldsymbol{\Delta}_{20}^{\mathrm{T}}, \cdots, \boldsymbol{\Delta}_{m0}^{\mathrm{T}})^{\mathrm{T}}$. 即约束条件(2.5)等价于问题(2.4)中的等式约束.

然而, 对于无参考中心的编队问题, 我们并没有假设在机器人队形中实际存在某个参考中心, 因此我们并没有 $\boldsymbol{p}_0$ 的坐标信息. 当机器人群构成的网络满足拓扑连通时, $\tilde{\boldsymbol{L}}(\boldsymbol{p} - \boldsymbol{\Delta}_0) = 0$ 等价于对于任意的 $i = 1, 2, \cdots, m$, 有

$$\sum_{j \in N_i} a_{ij}(\boldsymbol{p}_i - \boldsymbol{\Delta}_{i0} - \boldsymbol{p}_j + \boldsymbol{\Delta}_{j0}) = 0 \tag{2.7}$$

其中 N_i 表示机器人 i 的邻居指标集合, 如机器人 i 的邻居是机器人 $i-1$ 和 $i+1$, 那么 $N_i = \{i-1, i+1\}$.

由于 $\boldsymbol{\Delta}_{i0} - \boldsymbol{\Delta}_{j0} = \boldsymbol{\Delta}_{ij}$, 所以(2.7)式又可以写成

$$\sum_{j \in N_i} a_{ij}(\boldsymbol{p}_i - \boldsymbol{p}_j) = \sum_{j \in N_i} a_{ij} \boldsymbol{\Delta}_{ij} \quad (i = 1, 2, \cdots, m) \tag{2.8}$$

进而, 如果记 $\boldsymbol{q}_0 = (\sum_{j \in N_1} a_{1j} \boldsymbol{\Delta}_{1j}^{\mathrm{T}}, \sum_{j \in N_2} a_{2j} \boldsymbol{\Delta}_{2j}^{\mathrm{T}}, \cdots, \sum_{j \in N_m} a_{mj} \boldsymbol{\Delta}_{mj}^{\mathrm{T}})^{\mathrm{T}}$, 那么(2.8)式等价于

$$\tilde{\boldsymbol{L}}\boldsymbol{p} = \boldsymbol{q}_0$$

和上一节类似, 对于无参考中心的多机器人编队, 可以建立如下的优化问题:

$$\begin{aligned} \min_{\boldsymbol{p}} \quad & \frac{1}{2}\|\boldsymbol{p} - \boldsymbol{\xi}\|^2 \\ \text{s.t.} \quad & \begin{cases} \tilde{\boldsymbol{L}}\boldsymbol{p} = \boldsymbol{q}_0 \\ \boldsymbol{p} \in \Omega \end{cases} \end{aligned} \tag{2.9}$$

综合上述关于有参考中心和无参考中心的多机器人编队问题, 可以统一建立如下的约束优化模型:

$$\begin{aligned} \min_{\boldsymbol{p}} \quad & \frac{1}{2}\|\boldsymbol{p} - \boldsymbol{\xi}\|^2 \\ \text{s.t.} \quad & \begin{cases} \tilde{\boldsymbol{L}}\boldsymbol{p} = \boldsymbol{q} \\ \boldsymbol{p} \in \Omega \end{cases} \end{aligned} \tag{2.10}$$

其中 $\boldsymbol{q}$ 取值如下: 对于有现实参考中心的编队问题, 取(2.2)式中的 $\boldsymbol{q} = \tilde{\boldsymbol{L}}\boldsymbol{\Delta}_1$; 对于有虚拟参考中心的编队问题, 取(2.4)式中的 $\boldsymbol{q} = \tilde{\boldsymbol{L}}\boldsymbol{\Delta}_0$; 对于无参考中心的编队问题, 取(2.9)式中的 $\boldsymbol{q} = \boldsymbol{q}_0$.

接下来, 将进一步研究求解优化问题(2.10)的集中式算法和分布式算法.

2.3 集中式优化算法

这一节中,将介绍集中式算法如何求解优化问题(2.10).将基于梯度方法设计求解此类优化问题的迭代算法,同时证明该算法的收敛性. 根据鞍点定理1.6,可以得出如下关于问题(2.10)最优解的充分必要条件:

> **引理 2.2**
>
> $\boldsymbol{p}^* \in \mathbf{R}^{mn}$ 是优化问题(2.10)的最优解,当且仅当存在 $\boldsymbol{\beta}^* \in \mathbf{R}^{mn}$ 使得 $\boldsymbol{p}^*$ 和 $\boldsymbol{\beta}^*$ 满足如下的等式方程:
>
> $$\begin{cases} \boldsymbol{p}^* = \phi(\boldsymbol{p}^* - \sigma(\boldsymbol{p}^* - \boldsymbol{\xi} + \tilde{\boldsymbol{L}}\boldsymbol{\beta}^*)) \\ \tilde{\boldsymbol{L}}\boldsymbol{p}^* = \boldsymbol{q} \end{cases} \tag{2.11}$$
>
> 其中 $\sigma > 0$ 是常数,$\phi: \mathbf{R}^{mn} \to \Omega$ 是投影算子.

证明 问题(2.10)的拉格朗日函数可以表示为

$$\Phi(\boldsymbol{p}, \boldsymbol{\beta}) = \frac{1}{2}\|\boldsymbol{p} - \boldsymbol{\xi}\|^2 + \boldsymbol{\beta}^{\mathrm{T}}(\tilde{\boldsymbol{L}}\boldsymbol{p} - \boldsymbol{q})$$

其中 $\boldsymbol{\beta} \in \mathbf{R}^{mn}$ 是拉格朗日乘子. 根据鞍点定理1.6,$\boldsymbol{p}^*$ 是优化问题(2.10)的最优解,当且仅当 $(\boldsymbol{p}^*, \boldsymbol{\lambda}^*)$ 是拉格朗日函数的鞍点. 容易求得,函数 $\Phi(\boldsymbol{p}, \boldsymbol{\beta})$ 关于 $\boldsymbol{p}$ 的梯度 $\nabla\Phi(\boldsymbol{p}, \boldsymbol{\beta}) = \boldsymbol{p} - \boldsymbol{\xi} + \tilde{\boldsymbol{L}}\boldsymbol{\beta}$. 进而结合关于变分不等式问题的定理1.4,对任意 $\boldsymbol{p} \in \Omega$,$(\boldsymbol{p}^*, \boldsymbol{\beta}^*)$ 满足

$$(\boldsymbol{p} - \boldsymbol{p}^*)^{\mathrm{T}}(\boldsymbol{p}^* - \boldsymbol{\xi} + \tilde{\boldsymbol{L}}\boldsymbol{\beta}^*) \geqslant 0 \tag{2.12a}$$

$$\tilde{\boldsymbol{L}}\boldsymbol{p}^* = \boldsymbol{q} \tag{2.12b}$$

根据定理1.5,$\boldsymbol{p}^*$ 是变分不等式问题(2.12a)的解,当且仅当其满足如下的投影方程:

$$\boldsymbol{p}^* = \phi(\boldsymbol{p}^* - \sigma(\boldsymbol{p}^* - \boldsymbol{\xi} + \tilde{\boldsymbol{L}}\boldsymbol{\beta}^*))$$

其中 ϕ 是 $\mathbf{R}^{mn}$ 到 Ω 的投影算子,σ 是正常数. 证毕!

根据引理2.2,设计求解优化问题(2.10)的集中式算法的迭代公式为

$$\begin{cases} \boldsymbol{p}(k+1) = \phi(\boldsymbol{p}(k) - \sigma(\boldsymbol{p}(k) - \boldsymbol{\xi} + \boldsymbol{u}(k) + \tilde{\boldsymbol{L}}\boldsymbol{p}(k) - \boldsymbol{q})) \\ \boldsymbol{u}(k+1) = \boldsymbol{u}(k) + \tilde{\boldsymbol{L}}\boldsymbol{p}(k+1) - \boldsymbol{q} \end{cases} \tag{2.13}$$

接下来, 为了证明算法(2.13)的收敛性, 引入如下的迭代公式:

$$\begin{cases} \boldsymbol{p}(k+1) = \phi(\boldsymbol{p}(k) - \sigma(\boldsymbol{p}(k) - \boldsymbol{\xi} + \tilde{\boldsymbol{L}}(\boldsymbol{\beta}(k) + \boldsymbol{p}(k) - \boldsymbol{\mu}))) \\ \boldsymbol{\beta}(k+1) = \boldsymbol{\beta}(k) + \boldsymbol{p}(k+1) - \boldsymbol{\mu} \end{cases} \quad (2.14)$$

其中 $\boldsymbol{\mu}$ 满足 $\tilde{\boldsymbol{L}}\boldsymbol{\mu} = \boldsymbol{q}$.

注 对于问题(2.10), 一定存在 $\boldsymbol{\mu} \in \mathbf{R}^{mn}$ 使得 $\tilde{\boldsymbol{L}}\boldsymbol{\mu} = \boldsymbol{q}$. 事实上, 关于 $\boldsymbol{\mu}$ 的取值可以分为如下三种情况: 对于有现实参考中心的编队问题, 可以取(2.2)式中的 $\boldsymbol{\Delta}_1 = \boldsymbol{\mu}$; 对于有虚拟参考中心的编队问题, 可以取(2.4)式中的 $\boldsymbol{\Delta}_0 = \boldsymbol{\mu}$; 对于无参考中心的编队问题, 可以取(2.6)式中的 $\boldsymbol{\Delta}_0 = (\boldsymbol{\Delta}_{10}^{\mathrm{T}}, \boldsymbol{\Delta}_{20}^{\mathrm{T}}, \cdots, \boldsymbol{\Delta}_{m0}^{\mathrm{T}})^{\mathrm{T}}$ 使得 $\boldsymbol{\mu} = \boldsymbol{\Delta}_0$, 其满足 $\tilde{\boldsymbol{L}}\boldsymbol{\Delta}_0 = \boldsymbol{q}$.

为了简化起见, 以下的分析中将分别用下标形式的 $\boldsymbol{p}_k$ 和 $\boldsymbol{\beta}_k$ 代替 $\boldsymbol{p}(k)$ 和 $\boldsymbol{\beta}(k)$.

可以证明迭代公式(2.13)和(2.14)在一定条件下是等价的.

引理 2.3

对任意 $\boldsymbol{\beta}_0 \in \mathbf{R}^{mn}$, 如果初始值满足 $\boldsymbol{u}_0 = \tilde{\boldsymbol{L}}\boldsymbol{\beta}_0$, 那么迭代公式(2.13)和(2.14)是等价的.

证明 在公式(2.14)中, 令 $\boldsymbol{u}_k = \tilde{\boldsymbol{L}}\boldsymbol{\beta}_k$, 则很容易从公式(2.14)推出公式(2.13).

反之, 用数学归纳法证明其成立. 在公式(2.13)中, 因为 $\boldsymbol{u}_0 = \tilde{\boldsymbol{L}}\boldsymbol{\beta}_0$, 假设 $\boldsymbol{u}_k = \tilde{\boldsymbol{L}}\boldsymbol{\beta}_k$ 成立. 令 $\boldsymbol{\beta}_{k+1} = \boldsymbol{\beta}_k + \boldsymbol{p}_{k+1} - \boldsymbol{\mu}$. 根据(2.13)的第二个公式和 $\tilde{\boldsymbol{L}}\boldsymbol{\mu} = \boldsymbol{q}$, 可以得出

$$\begin{aligned} \boldsymbol{u}_{k+1} &= \boldsymbol{u}_k + \tilde{\boldsymbol{L}}\boldsymbol{p}_{k+1} - \boldsymbol{q} \\ &= \tilde{\boldsymbol{L}}\boldsymbol{\beta}_k + \tilde{\boldsymbol{L}}\boldsymbol{p}_{k+1} - \tilde{\boldsymbol{L}}\boldsymbol{\mu} \\ &= \tilde{\boldsymbol{L}}\boldsymbol{\beta}_{k+1} \end{aligned}$$

因此 $\boldsymbol{u}_k = \tilde{\boldsymbol{L}}\boldsymbol{\beta}_k$ 对任意 $k \in \{0,1,2,\cdots\}$ 总是成立. 即从公式(2.13)可以推出公式(2.14).

证毕!

接下来, 将研究算法(2.14)的收敛性. 设 $\boldsymbol{p}^* \in \Omega$ 是问题(2.10)的最优解, 根据引理2.2的结果, 存在 $\boldsymbol{\beta}^*$ 使得(2.11)中的等式成立. 定义如下的函数:

$$\begin{aligned} V_1(\boldsymbol{p}_k) &= \|\boldsymbol{p}_k - \boldsymbol{p}^*\|^2, \\ V_2(\boldsymbol{\beta}_k) &= (\boldsymbol{\beta}_k - \boldsymbol{\beta}^*)^{\mathrm{T}} \tilde{\boldsymbol{L}} (\boldsymbol{\beta}_k - \boldsymbol{\beta}^*) \end{aligned} \quad (2.15)$$

引理 2.4

根据 $V_1(\boldsymbol{p}_k)$ 和 $V_2(\boldsymbol{\beta}_k)$ 的定义, 对于迭代公式(2.14), 如下的不等式成立:

(i) $V_1(\boldsymbol{p}_{k+1}) - V_1(\boldsymbol{p}_k) \leqslant -(1-2\sigma)\|\boldsymbol{p}_{k+1} - \boldsymbol{p}_k\|^2$
$\quad - 2\sigma(\boldsymbol{p}_{k+1} - \boldsymbol{\mu})^{\mathrm{T}} \tilde{\boldsymbol{L}}(\boldsymbol{\beta}_k - \boldsymbol{\beta}^* + \boldsymbol{p}_k - \boldsymbol{\mu});$

(ii) $V_2(\boldsymbol{\beta}_{k+1}) - V_2(\boldsymbol{\beta}_k) = 2(\boldsymbol{p}_{k+1} - \boldsymbol{\mu})^{\mathrm{T}}\tilde{\boldsymbol{L}}(\boldsymbol{\beta}_k - \boldsymbol{\beta}^* + \boldsymbol{p}_k - \boldsymbol{\mu})$
$\qquad + (\boldsymbol{p}_{k+1} - \boldsymbol{p}_k)^{\mathrm{T}}\tilde{\boldsymbol{L}}(\boldsymbol{p}_{k+1} - \boldsymbol{p}_k)$
$\qquad - (\boldsymbol{p}_k - \boldsymbol{\mu})^{\mathrm{T}}\tilde{\boldsymbol{L}}(\boldsymbol{p}_k - \boldsymbol{\mu}).$

证明 (i) 设
$$\boldsymbol{\varphi}_k = \phi(\boldsymbol{p}_k - \sigma(\boldsymbol{p}_k - \boldsymbol{\xi} + \tilde{\boldsymbol{L}}(\boldsymbol{\beta}_k + \boldsymbol{p}_k - \boldsymbol{\mu})))$$

则 $\boldsymbol{p}_{k+1} = \boldsymbol{\varphi}_k$. 可以得到

$$\begin{aligned} &V_1(\boldsymbol{p}_{k+1}) - V_1(\boldsymbol{p}_k) \\ &= \|\boldsymbol{p}_{k+1} - \boldsymbol{p}^*\|^2 - \|\boldsymbol{p}_k - \boldsymbol{p}^*\|^2 \\ &= \|\boldsymbol{\varphi}_k - \boldsymbol{p}^*\|^2 - \|\boldsymbol{p}_k - \boldsymbol{p}^*\|^2 \\ &= (\boldsymbol{\varphi}_k - \boldsymbol{p}_k)^{\mathrm{T}}(\boldsymbol{\varphi}_k + \boldsymbol{p}_k - 2\boldsymbol{p}^*) \\ &= -\|\boldsymbol{\varphi}_k - \boldsymbol{p}_k\|^2 + 2(\boldsymbol{\varphi}_k - \boldsymbol{p}_k)^{\mathrm{T}}(\boldsymbol{\varphi}_k - \boldsymbol{p}^*) \\ &= -\|\boldsymbol{\varphi}_k - \boldsymbol{p}_k\|^2 + 2(\boldsymbol{\varphi}_k - \boldsymbol{p}_k + \sigma(\boldsymbol{p}_k - \boldsymbol{\xi} + \tilde{\boldsymbol{L}}(\boldsymbol{\beta}_k + \boldsymbol{p}_k - \boldsymbol{\mu})))^{\mathrm{T}}(\boldsymbol{\varphi}_k - \boldsymbol{p}^*) \\ &\quad - 2\sigma(\boldsymbol{p}_k - \boldsymbol{\xi} + \tilde{\boldsymbol{L}}(\boldsymbol{\beta}_k + \boldsymbol{p}_k - \boldsymbol{\mu}))^{\mathrm{T}}(\boldsymbol{\varphi}_k - \boldsymbol{p}^*) \end{aligned}$$

根据投影定理1.1, 可以得到

$$(\boldsymbol{\varphi}_k - \boldsymbol{p}_k + \sigma(\boldsymbol{p}_k - \boldsymbol{\xi} + \tilde{\boldsymbol{L}}(\boldsymbol{\beta}_k + \boldsymbol{p}_k - \boldsymbol{\mu})))^{\mathrm{T}}(\boldsymbol{\varphi}_k - \boldsymbol{p}^*) \leqslant 0$$

结合 $\boldsymbol{p}_{k+1} = \boldsymbol{\varphi}_k$, 有

$$\begin{aligned} &V_1(\boldsymbol{p}_{k+1}) - V_1(\boldsymbol{p}_k) \\ &\leqslant -\|\boldsymbol{p}_{k+1} - \boldsymbol{p}_k\|^2 - 2\sigma(\boldsymbol{p}_{k+1} - \boldsymbol{p}^*)^{\mathrm{T}}(\boldsymbol{p}_k - \boldsymbol{\xi} + \tilde{\boldsymbol{L}}(\boldsymbol{\beta}_k + \boldsymbol{p}_k - \boldsymbol{\mu})) \end{aligned} \quad (2.16)$$

设 $f(\boldsymbol{p}) = \|\boldsymbol{p} - \boldsymbol{\xi}\|^2/2$, 根据凸函数的性质 (定理1.2), 可以得到

$$f(\boldsymbol{p}_k) - f(\boldsymbol{p}^*) \leqslant (\boldsymbol{p}_k - \boldsymbol{p}^*)^{\mathrm{T}}(\boldsymbol{p}_k - \boldsymbol{\xi})$$
$$f(\boldsymbol{p}_{k+1}) - f(\boldsymbol{p}_k) \leqslant (\boldsymbol{p}_{k+1} - \boldsymbol{p}_k)^{\mathrm{T}}(\boldsymbol{p}_{k+1} - \boldsymbol{\xi})$$

进一步可以推出

$$\begin{aligned} &(\boldsymbol{p}_{k+1} - \boldsymbol{p}^*)^{\mathrm{T}}(\boldsymbol{p}_k - \boldsymbol{\xi}) \\ &= (\boldsymbol{p}_k - \boldsymbol{p}^*)^{\mathrm{T}}(\boldsymbol{p}_k - \boldsymbol{\xi}) + (\boldsymbol{p}_{k+1} - \boldsymbol{p}_k)^{\mathrm{T}}(\boldsymbol{p}_k - \boldsymbol{\xi}) \\ &\geqslant f(\boldsymbol{p}_k) - f(\boldsymbol{p}^*) + (\boldsymbol{p}_{k+1} - \boldsymbol{p}_k)^{\mathrm{T}}(\boldsymbol{p}_k - \boldsymbol{\xi}) \\ &= f(\boldsymbol{p}_k) - f(\boldsymbol{p}^*) + (\boldsymbol{p}_{k+1} - \boldsymbol{p}_k)^{\mathrm{T}}(\boldsymbol{p}_k - \boldsymbol{p}_{k+1}) + (\boldsymbol{p}_{k+1} - \boldsymbol{p}_k)^{\mathrm{T}}(\boldsymbol{p}_{k+1} - \boldsymbol{\xi}) \\ &\geqslant f(\boldsymbol{p}_{k+1}) - f(\boldsymbol{p}^*) - \|\boldsymbol{p}_{k+1} - \boldsymbol{p}_k\|^2 \end{aligned}$$

将上式代入(2.16)式, 可以得到

$$\begin{aligned}V_1(\boldsymbol{p}_{k+1}) - V_1(\boldsymbol{p}_k) \\ \leqslant -(1-2\sigma)\|\boldsymbol{p}_{k+1}-\boldsymbol{p}_k\|^2 + 2\sigma(f(\boldsymbol{p}^*)-f(\boldsymbol{p}_{k+1})) \\ -2\sigma(\boldsymbol{p}_{k+1}-\boldsymbol{p}^*)^{\mathrm{T}}\tilde{\boldsymbol{L}}(\boldsymbol{\beta}_k+\boldsymbol{p}_k-\boldsymbol{\mu})\end{aligned} \quad (2.17)$$

根据鞍点定理1.6, $(\boldsymbol{p}^*,\boldsymbol{\beta}^*)$ 是如下拉格朗日函数的鞍点:

$$\Phi(\boldsymbol{p},\boldsymbol{\beta}) = \frac{1}{2}\|\boldsymbol{p}-\boldsymbol{\xi}\|^2 + \boldsymbol{\beta}^{\mathrm{T}}(\tilde{\boldsymbol{L}}\boldsymbol{p}-\boldsymbol{q})$$

即 $(\boldsymbol{p}^*,\boldsymbol{\beta}^*)$ 满足

$$\Phi(\boldsymbol{p}^*,\boldsymbol{\beta}) \leqslant \Phi(\boldsymbol{p}^*,\boldsymbol{\beta}^*) \leqslant \Phi(\boldsymbol{p},\boldsymbol{\beta}^*)$$

因此, $(\boldsymbol{p}^*,\boldsymbol{\beta}^*)$ 是函数 $\Phi(\boldsymbol{p},\boldsymbol{\beta}^*)$ 关于变量 $\boldsymbol{p}$ 的最小值点. 根据定理1.4, $\boldsymbol{p}^*$ 是如下变分不等式的解:

$$(\boldsymbol{p}-\boldsymbol{p}^*)^{\mathrm{T}}(\boldsymbol{p}^*-\boldsymbol{\xi}+\tilde{\boldsymbol{L}}\boldsymbol{\beta}^*) \geqslant 0, \quad \forall \boldsymbol{p} \in \Omega$$

由于 $\boldsymbol{p}_{k+1} \in \Omega$, 所以

$$(\boldsymbol{p}_{k+1}-\boldsymbol{p}^*)^{\mathrm{T}}(\boldsymbol{p}^*-\boldsymbol{\xi}+\tilde{\boldsymbol{L}}\boldsymbol{\beta}^*) \geqslant 0 \quad (2.18)$$

进而根据凸函数的性质, 有

$$f(\boldsymbol{p}_{k+1}) - f(\boldsymbol{p}^*) \geqslant (\boldsymbol{p}_{k+1}-\boldsymbol{p}^*)^{\mathrm{T}}(\boldsymbol{p}^*-\boldsymbol{\xi})$$

结合不等式(2.18), 得到

$$f(\boldsymbol{p}_{k+1}) - f(\boldsymbol{p}^*) + (\boldsymbol{p}_{k+1}-\boldsymbol{p}^*)^{\mathrm{T}}\tilde{\boldsymbol{L}}\boldsymbol{\beta}^* \geqslant 0$$

进而结合不等式(2.17)可以推出

$$\begin{aligned}V_1(\boldsymbol{p}_{k+1}) - V_1(\boldsymbol{p}_k) \\ \leqslant -(1-2\sigma)\|\boldsymbol{p}_{k+1}-\boldsymbol{p}_k\|^2 - 2\sigma(\boldsymbol{p}_{k+1}-\boldsymbol{p}^*)^{\mathrm{T}}\tilde{\boldsymbol{L}}(\boldsymbol{\beta}_k-\boldsymbol{\beta}^*+\boldsymbol{p}_k-\boldsymbol{\mu})\end{aligned}$$

注意到 $\tilde{\boldsymbol{L}}\boldsymbol{p}^* = \tilde{\boldsymbol{L}}\boldsymbol{\mu} = \boldsymbol{q}$, 将其代入上一个不等式, 得到

$$\begin{aligned}V_1(\boldsymbol{p}_{k+1}) - V_1(\boldsymbol{p}_k) \\ \leqslant -(1-2\sigma)\|\boldsymbol{p}_{k+1}-\boldsymbol{p}_k\|^2 - 2\sigma(\tilde{\boldsymbol{L}}\boldsymbol{p}_{k+1}-\boldsymbol{q})^{\mathrm{T}}(\boldsymbol{\beta}_k-\boldsymbol{\beta}^*+\boldsymbol{p}_k-\boldsymbol{\mu}) \\ = -(1-2\sigma)\|\boldsymbol{p}_{k+1}-\boldsymbol{p}_k\|^2 - 2\sigma(\boldsymbol{p}_{k+1}-\boldsymbol{\mu})^{\mathrm{T}}\tilde{\boldsymbol{L}}(\boldsymbol{\beta}_k-\boldsymbol{\beta}^*+\boldsymbol{p}_k-\boldsymbol{\mu})\end{aligned}$$

(ii) 因为 $\boldsymbol{\beta}_{k+1} = \boldsymbol{\beta}_k + \boldsymbol{p}_{k+1} - \boldsymbol{\mu}$, 所以

$$\begin{aligned}
&V_2(\boldsymbol{\beta}_{k+1}) - V_2(\boldsymbol{\beta}_k) \\
&= (\boldsymbol{\beta}_{k+1} - \boldsymbol{\beta}^*)^{\mathrm{T}} \tilde{\boldsymbol{L}} (\boldsymbol{\beta}_{k+1} - \boldsymbol{\beta}^*) - (\boldsymbol{\beta}_k - \boldsymbol{\beta}^*)^{\mathrm{T}} \tilde{\boldsymbol{L}} (\boldsymbol{\beta}_k - \boldsymbol{\beta}^*) \\
&= (\boldsymbol{\beta}_k - \boldsymbol{\beta}^* + \boldsymbol{p}_{k+1} - \boldsymbol{\mu})^{\mathrm{T}} \tilde{\boldsymbol{L}} (\boldsymbol{\beta}_k - \boldsymbol{\beta}^* + \boldsymbol{p}_{k+1} - \boldsymbol{\mu}) - (\boldsymbol{\beta}_k - \boldsymbol{\beta}^*)^{\mathrm{T}} \tilde{\boldsymbol{L}} (\boldsymbol{\beta}_k - \boldsymbol{\beta}^*) \\
&= 2(\boldsymbol{p}_{k+1} - \boldsymbol{\mu})^{\mathrm{T}} \tilde{\boldsymbol{L}} (\boldsymbol{\beta}_k - \boldsymbol{\beta}^*) + (\boldsymbol{p}_{k+1} - \boldsymbol{\mu})^{\mathrm{T}} \tilde{\boldsymbol{L}} (\boldsymbol{p}_{k+1} - \boldsymbol{\mu}) \\
&= 2(\boldsymbol{p}_{k+1} - \boldsymbol{\mu})^{\mathrm{T}} \tilde{\boldsymbol{L}} (\boldsymbol{\beta}_k - \boldsymbol{\beta}^*) + (\boldsymbol{p}_{k+1} - \boldsymbol{p}_k)^{\mathrm{T}} \tilde{\boldsymbol{L}} (\boldsymbol{p}_{k+1} - \boldsymbol{p}_k) \\
&\quad + 2(\boldsymbol{p}_{k+1} - \boldsymbol{\mu})^{\mathrm{T}} \tilde{\boldsymbol{L}} (\boldsymbol{p}_k - \boldsymbol{\mu}) - (\boldsymbol{p}_k - \boldsymbol{\mu})^{\mathrm{T}} \tilde{\boldsymbol{L}} (\boldsymbol{p}_k - \boldsymbol{\mu}) \\
&= 2(\boldsymbol{p}_{k+1} - \boldsymbol{\mu})^{\mathrm{T}} \tilde{\boldsymbol{L}} (\boldsymbol{\beta}_k - \boldsymbol{\beta}^* + \boldsymbol{p}_k - \boldsymbol{\mu}) + (\boldsymbol{p}_{k+1} - \boldsymbol{p}_k)^{\mathrm{T}} \tilde{\boldsymbol{L}} (\boldsymbol{p}_{k+1} - \boldsymbol{p}_k) \\
&\quad - (\boldsymbol{p}_k - \boldsymbol{\mu})^{\mathrm{T}} \tilde{\boldsymbol{L}} (\boldsymbol{p}_k - \boldsymbol{\mu})
\end{aligned}$$

证毕!

接下来, 给出迭代公式(2.14)收敛的主要结论.

定理 2.1

对于任意给定的初始值 $\boldsymbol{p}(0)$ 和 $\boldsymbol{\beta}(0)$, 如果参数 σ 满足

$$\sigma < \frac{1}{2 + \lambda_{\max}(\tilde{\boldsymbol{L}})}$$

那么迭代公式(2.14)中的 $\boldsymbol{p}(k)$ 能够全局收敛到问题(2.10)的最优解, 其中 $\lambda_{\max}$ 表示矩阵的最大特征值.

证明 设 $\boldsymbol{p}^*$ 是问题(2.10)的最优解. 根据引理(2.2), 存在 $\boldsymbol{\beta}^* \in \mathbf{R}^{mn}$ 使得(2.11)中的等式成立.

构造如下的李雅普诺夫函数 (附录 B):

$$V(\boldsymbol{p}_k, \boldsymbol{\beta}_k) = V_1(\boldsymbol{p}_k) + \sigma V_2(\boldsymbol{\beta}_k) \tag{2.19}$$

其中 $V_1(\boldsymbol{p}_k)$ 和 $V_2(\boldsymbol{\beta}_k)$ 如式(2.15)中的定义.

根据引理2.4的结论, 可以得到

$$\begin{aligned}
&V(\boldsymbol{p}_{k+1}, \boldsymbol{\beta}_{k+1}) - V(\boldsymbol{p}_k, \boldsymbol{\beta}_k) \\
&= V_1(\boldsymbol{p}_{k+1}) - V_1(\boldsymbol{p}_k) + \sigma(V_2(\boldsymbol{\beta}_{k+1}) - V_2(\boldsymbol{\beta}_k)) \\
&\leqslant -(1 - 2\sigma)\|\boldsymbol{p}_{k+1} - \boldsymbol{p}_k\|^2 + \sigma(\boldsymbol{p}_{k+1} - \boldsymbol{p}_k)^{\mathrm{T}} \tilde{\boldsymbol{L}} (\boldsymbol{p}_{k+1} - \boldsymbol{p}_k) - \sigma(\boldsymbol{p}_k - \boldsymbol{\mu})^{\mathrm{T}} \tilde{\boldsymbol{L}} (\boldsymbol{p}_k - \boldsymbol{\mu}) \\
&= -(\boldsymbol{p}_{k+1} - \boldsymbol{p}_k)^{\mathrm{T}} ((1 - 2\sigma)\boldsymbol{I} - \sigma\tilde{\boldsymbol{L}})(\boldsymbol{p}_{k+1} - \boldsymbol{p}_k) - \sigma(\boldsymbol{p}_k - \boldsymbol{\mu})^{\mathrm{T}} \tilde{\boldsymbol{L}} (\boldsymbol{p}_k - \boldsymbol{\mu})
\end{aligned} \tag{2.20}$$

其中 $\boldsymbol{I}$ 是单位矩阵.

如果 $\sigma < 1/(2+\lambda_{\max}(\tilde{\boldsymbol{L}}))$,那么矩阵 $(1-2\sigma)\boldsymbol{I}-\sigma\tilde{\boldsymbol{L}}$ 是正定的. 可以得到 $V(\boldsymbol{p}_{k+1},\boldsymbol{\beta}_{k+1})-V(\boldsymbol{p}_k,\boldsymbol{\beta}_k)\leqslant 0$.

根据拉萨尔 (LaSalle) 不变性原理 (附录 C), $\boldsymbol{p}_k$ 收敛到如下集合的最大不变集:

$$\Xi = \{\boldsymbol{p}_k \in \mathbf{R}^{mn} | V(\boldsymbol{p}_{k+1},\boldsymbol{\beta}_{k+1}) - V(\boldsymbol{p}_k,\boldsymbol{\beta}_k) = 0\}$$

根据式(2.20),如果 $V(\boldsymbol{p}_{k+1},\boldsymbol{\beta}_{k+1})-V(\boldsymbol{p}_k,\boldsymbol{\beta}_k)=0$,我们可以得到 $\boldsymbol{p}_{k+1}=\boldsymbol{p}_k$, $\tilde{\boldsymbol{L}}(\boldsymbol{p}_k-\boldsymbol{\mu})=0$. 即 $\boldsymbol{p}_k$ 满足方程

$$\begin{cases} \boldsymbol{p}_k = \phi(\boldsymbol{p}_k - \sigma(\boldsymbol{p}_k - \boldsymbol{\xi} + \tilde{\boldsymbol{L}}\boldsymbol{\beta}_k)) \\ \tilde{\boldsymbol{L}}\boldsymbol{p}_k = \boldsymbol{q} \end{cases}$$

其中第二个等式成立用到了 $\tilde{\boldsymbol{L}}\boldsymbol{\mu}=\boldsymbol{q}$. 也就是说, $\boldsymbol{p}_k$ 满足(2.11)中的等式. 因此, $\boldsymbol{p}_k$ 是问题(2.10)的最优解. 结合上述结论, $\boldsymbol{p}_k$ 将收敛到问题(2.10)的最优解集.

根据 $V(\boldsymbol{p}_k,\boldsymbol{\beta}_k)$ 的表达式和不等式(2.20),有 $V(\boldsymbol{p}_k,\boldsymbol{\beta}_k) \leqslant V(\boldsymbol{p}_0,\boldsymbol{\beta}_0)$. 进而 $\boldsymbol{p}_k$ 和 $\tilde{\boldsymbol{L}}\boldsymbol{\beta}_k$ 是有界的. 因此存在递增序列 $\{k_l\}_{l=1}^{\infty}$ 和极限点 $\boldsymbol{p}'$ 和 $\boldsymbol{\beta}'_L$ 使得 $\lim_{l\to\infty}\boldsymbol{p}_{k_l}=\boldsymbol{p}'$ 和 $\lim_{l\to\infty}\tilde{\boldsymbol{L}}\boldsymbol{\beta}_{k_l}=\boldsymbol{\beta}'_L$,其中 $\boldsymbol{p}'$ 是问题(2.10)的最优解. 根据 $\lim_{l\to\infty}\tilde{\boldsymbol{L}}\boldsymbol{\beta}_{k_l}=\boldsymbol{\beta}'_L$,以及线性空间 $\tilde{\boldsymbol{L}}\boldsymbol{\beta}$ 的完备性,存在 $\boldsymbol{\beta}' \in \mathbf{R}^{mn}$ 使得 $\boldsymbol{L}\boldsymbol{\beta}'=\boldsymbol{\beta}'_L$. 因此 $(\boldsymbol{p}',\boldsymbol{\beta}')$ 满足(2.11)中的等式.

定义另一个函数 $V' = \|\boldsymbol{p}_k - \boldsymbol{p}'\|^2 + \sigma(\boldsymbol{\beta}_k - \boldsymbol{\beta}')^{\mathrm{T}}\tilde{\boldsymbol{L}}(\boldsymbol{\beta}_k - \boldsymbol{\beta}')$,其形式上就是将公式(2.19)中的 $\boldsymbol{p}^*$ 和 $\boldsymbol{\beta}^*$ 换成 $\boldsymbol{p}'$ 和 $\boldsymbol{\beta}'$. 一方面,类似于上述分析,得到 $V'(\boldsymbol{p}_k,\boldsymbol{\beta}_k) \leqslant V'(\boldsymbol{p}_{k-1},\boldsymbol{\beta}_{k-1})$. 另一方面,对于任意 k,存在 k_l 使得 $V'(\boldsymbol{p}_k,\boldsymbol{\beta}_k) \leqslant V'(\boldsymbol{p}_{k_l},\boldsymbol{\beta}_{k_l})$. 令 $l\to\infty$,则 $k\to\infty$,进而 $\lim_{k\to\infty}V'(\boldsymbol{p}_k,\boldsymbol{\beta}_k) = \lim_{l\to\infty}V'(\boldsymbol{p}_{k_l},\boldsymbol{\beta}_{k_l}) = 0$. 由于 $V'(\boldsymbol{p}_k,\boldsymbol{\beta}_k) \geqslant \|\boldsymbol{p}_k-\boldsymbol{p}'\|^2$,有 $\lim_{k\to\infty}\boldsymbol{p}_k = \boldsymbol{p}'$,即 $\boldsymbol{p}_k$ 全局收敛到问题(2.10)的最优解. 证毕!

2.4 分布式优化算法

对上述优化算法(2.14)的实现,如果是在单台处理器上进行计算,那么该算法的实现是集中式的. 事实上, 也可以采用分布式的计算方式进行处理. 如果将算法(2.14)写成分量的形式,那么可以清楚地看出该算法的分布式处理方式.

对于机器人 i,其迭代公式为

$$\begin{cases} \boldsymbol{p}_i(k+1) = \phi_i(\boldsymbol{p}_i(k) - \sigma(\boldsymbol{p}_i(k) - \boldsymbol{\xi}_i + \boldsymbol{u}_i(k) + \sum_{j\in N_i} a_{ij}(\boldsymbol{p}_i(k) - \boldsymbol{p}_j(k)) - \boldsymbol{q}_i)) \\ \boldsymbol{u}_i(k+1) = \boldsymbol{u}_i(k) + \sum_{j\in N_i} a_{ij}(\boldsymbol{p}_i(k+1) - \boldsymbol{p}_j(k+1)) - \boldsymbol{q}_i \end{cases} \quad (2.21)$$

其中关于 $\boldsymbol{q}_i$ 的取值,对于优化问题(2.2), $\boldsymbol{q}_i = \sum_{j\in N_i} a_{ij}(\boldsymbol{\Delta}_{i1}^{\mathrm{T}} - \boldsymbol{\Delta}_{j1}^{\mathrm{T}})$;对于优化问题(2.4),

$q_i = \sum_{j \in N_i} a_{ij}(\mathbf{\Delta}_{i0}^{\mathrm{T}} - \mathbf{\Delta}_{j0}^{\mathrm{T}})$; 对于优化问题(2.9)，$q_i = \sum_{j \in N_i} a_{ij} \mathbf{\Delta}_{ij}^{\mathrm{T}}$.

接下来，具体比较上述三种情况在算法求解中的异同点:
- 对于优化问题(2.2)，机器人 i 需要用到自身的状态信息 $\mathbf{p}_i$ 和 $\mathbf{u}_i$，自身的队形信息 $\mathbf{\Delta}_{i1}$，以及与其进行通信的邻居状态信息 $\mathbf{p}_j$ 和邻居队形信息 $\mathbf{\Delta}_{j1}$;
- 对于优化问题(2.4)，机器人 i 需要用到自身的状态信息 $\mathbf{p}_i$ 和 $\mathbf{u}_i$，自身的队形信息 $\mathbf{\Delta}_{i0}$，以及与其进行通信的邻居状态信息 $\mathbf{p}_j$ 和邻居队形信息 $\mathbf{\Delta}_{j0}$;
- 对于优化问题(2.9)，机器人 i 需要用到自身的状态信息 $\mathbf{p}_i$ 和 $\mathbf{u}_i$，与其进行通信的邻居状态信息 $\mathbf{p}_j$，以及其与邻居的相对位置信息 $\mathbf{\Delta}_{ij}$.

从上述比较可以看出，对优化问题(2.2)和(2.4)的求解算法，需要知道机器人相对于参考中心的位置信息，这可以看作是全局的绝对位置信息. 然而，对优化问题(2.9)的求解算法，只需要知道机器人和邻居的相对位置信息，这可以看作是局部的相对位置信息. 因此，优化问题(2.9)对多机器人编队具有更大的灵活性，也就是对于无参考中心的优化问题，其对应的分布式优化算法更具有分布式特点，在机器人通信拓扑和空间拓扑上保持了一致.

练 习

假设有 4 个机器人进行编队，对于分布式优化算法(2.21)，分别对于图2.4的星形和环形通信拓扑结构写出每个机器人对应的迭代算法.

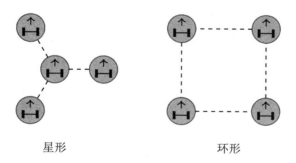

星形　　　　　　环形

图 2.4　四个机器人形成的两个通信拓扑结构

第 3 章　基于几何相似性的多机器人编队

上一章中介绍的多机器人编队问题中, 要求机器人间的相对位置是固定不变的, 然而这种方法在进行多机器人编队时灵活性不高. 本章将介绍基于几何相似性的多机器人编队问题, 将首先建立多机器人编队问题的优化模型, 进而设计相应的优化求解算法.

3.1　向量的几何变换

在解析几何的意义下, 平面上的点 A 可以看作起点为原点, 终点为 A 的向量. 向量的几何变换主要包括三种: 旋转、伸缩和平移. 下面分别以二维向量为例介绍这三种变换.

3.1.1　旋转变换

假设有二维平面上的向量 A, 其坐标为 (x,y), 如图3.1所示.

对平面上的向量 $A=(x,y)^{\mathrm{T}}$ 进行旋转变换, 即将其旋转一个角度 θ, 相当于对其左乘如下的旋转变换矩阵:

$$R=\begin{pmatrix} \cos\theta & -\sin\theta \\ \sin\theta & \cos\theta \end{pmatrix} \tag{3.1}$$

将图3.1的向量 A 逆时针旋转角度 $\theta=30°$, 可以得到向量 B.

3.1.2　伸缩变换

对平面上的向量 $A=(x,y)^{\mathrm{T}}$ 进行伸缩变换, 即向量的角度不变, 将其长度变换为原来的 r 倍, 相当于用常数 r 乘以此向量, 即得到新的向量为 rA. 将图3.1的向量 A 缩短为原来的 0.5, 可以得到向量 C.

3.1.3　平移变换

对平面上的向量 $A=(x,y)^{\mathrm{T}}$ 进行平移变换, 即向量向左/右或者向上/下进行平移, 相当于对向量 A 的坐标 x 和 y 分别加上对应的平移量. 如果平移量为 $d=(\bar{x},\bar{y})^{\mathrm{T}}$, 则

新的向量为 $\boldsymbol{A}+\boldsymbol{d}$. 将图3.1中的向量 $\boldsymbol{A}$ 向左平移 1.5, 再向上平移 0.5, 可以得到向量 $\boldsymbol{D}$.

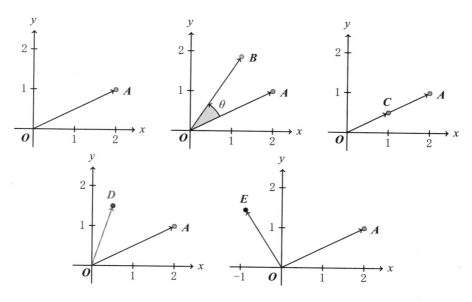

图 3.1　五个图形分别对应: 原始向量 $\boldsymbol{A}$; 将向量 $\boldsymbol{A}$ 进行旋转变换, 得到向量 $\boldsymbol{B}$; 将向量 $\boldsymbol{A}$ 进行伸缩变换, 得到向量 $\boldsymbol{C}$; 将向量 $\boldsymbol{A}$ 进行平移变化, 得到向量 $\boldsymbol{D}$; 将向量 $\boldsymbol{A}$ 进行混合变换, 得到向量 $\boldsymbol{E}$

3.1.4　混合变换

混合变换是指包含至少两种上述的变换. 设向量 $\boldsymbol{A}=(x,y)^{\mathrm{T}}$, 则将此向量同时经过旋转—伸缩—平移变换后得到的新向量可以表示为

$$\boldsymbol{z}=r\begin{pmatrix}\cos\theta & -\sin\theta \\ \sin\theta & \cos\theta\end{pmatrix}\begin{pmatrix}x \\ y\end{pmatrix}+\begin{pmatrix}\bar{x} \\ \bar{y}\end{pmatrix}$$

或简单表示为 $\boldsymbol{z}=r\boldsymbol{R}\boldsymbol{A}+\boldsymbol{d}$.

图3.1中, 将向量 $\boldsymbol{A}$ 逆时针旋转 $\theta=30°$, 缩短为原来的 $r=0.5$, 再平移 $\boldsymbol{d}=(-1.5,0.5)^{\mathrm{T}}$, 可以得到向量 $\boldsymbol{E}$.

3.2　实现多机器人编队的约束处理

基于几何相似性的队形编队, 机器人之间的相对位置是可以变化的, 但是整个机器人群的队形需要满足几何相似的特性. 因此, 首先给出队形等价类的定义.

> **定义 3.1** (队形等价类)
>
> 设 m 个机器人具有队形 S，这些机器人组成的队形等价类，是指由 m 个两两不重合点构成的具有一定几何形状的图标，这些图标在旋转、伸缩和平移变换下具有和 S 相同的几何形状，队形等价类用符号表示为 $[S]$.

考虑二维平面上由 m 个机器人形成队形的编队任务，以实现机器人群与给定队形具有相同的几何形状，其中**形状图标**(队形)在队形等价类的意义下，由 m 个点构成，记为 $\{s_1, s_2, \cdots, s_m\}$，每个 $s_i \in \mathbf{R}^2$. 机器人群的队形等价类可以表示为

$$[S] = \{r\bm{RS} + \bm{d}\mathbf{1}_m^{\mathrm{T}} | r > 0, \bm{R} \in SO(2), \bm{d} \in \mathbf{R}^2\} \tag{3.2}$$

其中 $r > 0$ 是伸缩因子，$\bm{R} \in SO(2)$ 是二维平面上的旋转矩阵 (如式(3.1)所示)，$\bm{d}$ 是平移向量，$\mathbf{1}_m$ 是元素全为 1 的 m 维向量.

用 $\bm{s}_i = (s_i^x, s_i^y)^{\mathrm{T}}$ 表示第 i 个机器人在队形等价类中所对应的坐标，$\bm{p} = (p_i^x, p_i^y)^{\mathrm{T}}$ 表示第 i 个机器人在真实二维坐标系下的目标位置，即是需要确定的未知变量.

3.2.1 队形约束的简化处理

为了方便理论分析，接下来将队形 S 进行简化处理[①]. 假设队形 S 中的 s_1 在原点位置，即 $s_1 = (0,0)^{\mathrm{T}}$；s_2 在 x 轴上，即 $s_2^y = 0$. 如图3.2所示，我们考虑三个机器人的队形，其他多个机器人的情况类似处理. 图中 $\bm{p}_1$ 和 $\bm{p}_2$ 可以处于平面上的任意位置，而 $\bm{p}_i$ $(i = 3, 4, \cdots, m)$ 的位置需要由队形 S 确定. 根据式(3.2)，$\bm{p}_1, \bm{p}_i$ 和 $\bm{s}_i$ $(i = 2, 3, \cdots, m)$ 应满足如下关系：

$$\bm{p}_i = r\bm{R}\bm{s}_i + \bm{p}_1 \quad (i = 2, 3, \cdots, m) \tag{3.3}$$

其中伸缩因子 r 和公式(3.1)中的旋转角通过如下公式计算：

$$r = \frac{\|\bm{p}_2 - \bm{p}_1\|}{\|\bm{s}_2\|}, \quad \theta = \arctan\frac{p_2^y - p_1^y}{p_2^x - p_1^x}$$

进而，式(3.3)可以写成坐标的形式

$$\begin{cases} p_i^x = r(s_i^x \cos\theta - s_i^y \sin\theta) + p_1^x \\ p_i^y = r(s_i^x \sin\theta + s_i^y \cos\theta) + p_1^y \end{cases} \quad (i = 2, 3, \cdots, m) \tag{3.4}$$

因为 $\|\bm{p}_2 - \bm{p}_1\|\cos\theta = p_2^x - p_1^x$ 和 $\|\bm{p}_2 - \bm{p}_1\|\sin\theta = p_2^y - p_1^y$，将其代入(3.4)，得到

[①] 这一部分的详细介绍可以参考文献 [4].

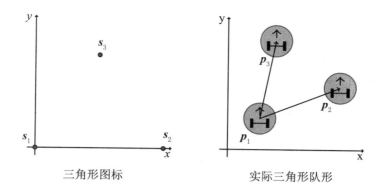

图 3.2 机器人编队的形状图标和实际队形

$$\begin{cases} \|\boldsymbol{s}_2\|(p_i^x - p_1^x) = (s_i^x, -s_i^y)(\boldsymbol{p}_2 - \boldsymbol{p}_1) \\ \|\boldsymbol{s}_2\|(p_i^y - p_1^y) = (s_i^y, s_i^x)(\boldsymbol{p}_2 - \boldsymbol{p}_1) \end{cases} \quad (i = 3, 4, \cdots, m) \tag{3.5}$$

注 公式(3.5)给出了机器人队形的约束条件, 其中没有考虑像(3.4)式中 $i=2$ 的约束条件. 这是因为对于二维平面上的编队问题, 第 1 和第 2 号机器人可以取得任意位置, 因此, 在约束条件(3.5)中无需考虑这两个机器人的位置约束.

公式(3.5)给出了编队问题所对应的 $2(m-2)$ 个等式约束, 可以将其写成向量和矩阵的形式, 以方便后续的分析. 记

$$\boldsymbol{M} = \begin{pmatrix} \|\boldsymbol{s}_2\| & 0 \\ 0 & \|\boldsymbol{s}_2\| \end{pmatrix}, \quad \boldsymbol{M}_i = \begin{pmatrix} s_i^x & -s_i^y \\ s_i^y & s_i^x \end{pmatrix}, \quad \boldsymbol{O} = \begin{pmatrix} 0 & 0 \\ 0 & 0 \end{pmatrix}$$

则式(3.5)可以等价地表示成

$$\boldsymbol{M}(\boldsymbol{p}_i - \boldsymbol{p}_1) = \boldsymbol{M}_i(\boldsymbol{p}_2 - \boldsymbol{p}_1) \quad (i = 3, 4, \cdots, m) \tag{3.6}$$

进而, 令

$$\boldsymbol{A}_1 = \begin{pmatrix} \boldsymbol{M}_3 - \boldsymbol{M} & -\boldsymbol{M}_3 & \boldsymbol{M} & \boldsymbol{O} & \cdots & \boldsymbol{O} \\ \boldsymbol{M}_4 - \boldsymbol{M} & -\boldsymbol{M}_4 & \boldsymbol{O} & \boldsymbol{M} & \cdots & \boldsymbol{O} \\ \vdots & \vdots & \vdots & \vdots & & \vdots \\ \boldsymbol{M}_m - \boldsymbol{M} & -\boldsymbol{M}_m & \boldsymbol{O} & \boldsymbol{O} & \cdots & \boldsymbol{M} \end{pmatrix}, \quad \boldsymbol{p} = \begin{pmatrix} \boldsymbol{p}_1 \\ \boldsymbol{p}_2 \\ \vdots \\ \boldsymbol{p}_m \end{pmatrix}$$

则式(3.6)可以写成如下的整体约束形式:

$$\boldsymbol{A}_1 \boldsymbol{p} = \boldsymbol{0}_{2(m-2)} \tag{3.7}$$

其中 $\boldsymbol{0}_{2(m-2)}$ 是所有元素都为 0 的 $2(m-2)$ 维向量.

3.2.2 带锚点的队形约束

接下来,考虑在队形约束条件(3.7)中加入锚点约束[①]. 所谓锚点,是指在队形中有一个或者多个机器人需要满足绝对或者相对的位置约束. 例如,要求第 1 号机器人需要在坐标原点,则约束条件为 $p_1 = (0,0)^{\mathrm{T}}$;如果要求第 1 和第 2 号机器人的相对位置在直线 $y = x$ 上,则约束条件为 $p_1^y - p_2^y = p_1^x - p_2^x$,即 $(1,-1)(p_1 - p_2) = 0$.

首先考虑加入一个锚点的情况. 不失一般性,将第一个机器人设置为固定锚点. 因此需要加入如下的等式约束:

$$\begin{pmatrix} I_2 & O & \cdots & O \end{pmatrix} p = b_1$$

其中 I_2 是 2 维单位矩阵,$b_1 \in \mathbf{R}^2$ 是一个固定点. 设 $B_1 = (I_2, O, \cdots, O)$,可以将式(3.7)中的等式约束扩展为

$$\begin{pmatrix} A_1 \\ B_1 \end{pmatrix} p = \begin{pmatrix} \mathbf{0}_{2(m-2)} \\ b_1 \end{pmatrix} \tag{3.8}$$

因为现考虑的是二维平面上的编队问题,因此可以再加入一个锚点. 如果添加 $q_2 = b_2$ 作为第二个锚点,则可以加入如下两种情况的等式约束:一种是固定锚点

$$\begin{pmatrix} I_n & O & O & \cdots & O \\ O & I_n & O & \cdots & O \end{pmatrix} p = \begin{pmatrix} b_1 \\ b_2 \end{pmatrix} \tag{3.9}$$

另一种是自由锚点

$$\begin{pmatrix} I_n & -I_n & O & \cdots & O \end{pmatrix} p = b_3 \tag{3.10}$$

其中 $b_2, b_3 \in \mathbf{R}^2$,$b_3$ 给出了第 1 和第 2 号机器人的相对位置,即 $p_1 - p_2 = b_3$.

设 B_2 和 B_3 分别表示式(3.9)和式(3.10)中等号左边的两个矩阵,则可以将式(3.7)中的等式约束扩展为

$$\begin{pmatrix} A_1 \\ B_2 \end{pmatrix} p = \begin{pmatrix} \mathbf{0}_{2(m-2)} \\ b_1 \\ b_2 \end{pmatrix}, \quad \begin{pmatrix} A_1 \\ B_3 \end{pmatrix} p = \begin{pmatrix} \mathbf{0}_{2(m-2)} \\ b_3 \end{pmatrix} \tag{3.11}$$

其中第一个约束预设了 p_1 和 p_2 在编队中的绝对位置分别为 b_1 和 b_2,而第二个约束表示 p_1 和 p_2 空间位置不是固定的,但是它们之间的相对位置受 $p_1 - p_2 = b_3$ 的限制.

[①] 这一部分的详细介绍可以参考文献 [8].

3.3 多机器人编队的优化问题

上一节中给出了多个机器人要形成一个给定的队形 S 所需要满足的约束条件，也就是等式(3.7)、(3.8)和(3.11)所给出的约束条件．这一节将研究多机器人编队中的优化目标．

假设 m 个机器人的初始位置为 $\xi_1, \xi_2, \cdots, \xi_m$，形成队形的目标位置为 $p_1, p_2, \cdots, p_m$．对于机器人 i，其初始位置到队形目标位置的移动距离为 $\|p_i - \xi_i\|$．对于 m 个机器人，总的移动距离为 $\sum_{i=1}^{m} \|p_i - \xi_i\|$．为了分析和算法实现上的方便，同样采用范数平方作为目标函数，其形式如下：

$$\frac{1}{2} \sum_{i=1}^{m} \|p_i - \xi_i\|^2 \tag{3.12}$$

此外，和前一章类似，为了限制机器人在一个有限的空间范围内形成队形，现给每个机器人限制一个范围 $p_i \in \Omega_i$，其中 Ω_i 是机器人 i 在平面上可移动的范围．

综合上述关于多机器人编队的目标函数和约束条件，得出如下的优化问题：

$$\min_{p} \quad \frac{1}{2}\|p - \xi\|^2 \\ \text{s.t.} \begin{cases} Ap = b \\ p \in \Omega \end{cases} \tag{3.13}$$

其中 $p = (p_1^T, p_2^T, \cdots, p_m^T)^T$ 表示 m 个机器人期望到达的目标队形的位置，$\xi = (\xi_1^T, \xi_2^T, \cdots, \xi_m^T)^T$ 表示 m 个机器人的初始位置，A 和 b 分别表示等式约束(3.7)、(3.8)和(3.11)中等号左右两边的矩阵和向量，$\Omega = \prod_{i=1}^{m} \Omega_i$ 表示 Ω_i 的笛卡儿乘积．

3.4 集中式优化算法

为了求解优化问题(3.13)，接下来将设计相应的求解算法．根据鞍点定理1.6，可以得出如下关于问题(3.13)最优解的充分必要条件：

引理 3.1

$p^* \in \mathbf{R}^{2m}$ 是优化问题(3.13)的最优解，当且仅当存在 $y^* \in \mathbf{R}^l$（l 是等式约束中矩阵 A 的行数）使得 p^* 和 y^* 满足如下的等式方程：

$$\begin{cases} p^* = \phi(p^* - \sigma(p^* - \xi) + A^T y^*) \\ Ap^* = b \end{cases} \tag{3.14}$$

其中 $\sigma > 0$ 是常数，$\phi: \mathbf{R}^{2m} \to \Omega$ 是投影算子.

注 引理3.1的证明类似引理2.2的证明，将其留作练习.

根据引理3.1，求解优化问题(3.13)的集中式算法可以表示为

$$\begin{cases} \boldsymbol{p}(k+1) = \phi(\boldsymbol{p}(k) - \sigma(\boldsymbol{p}(k) - \boldsymbol{\xi} + \boldsymbol{A}^{\mathrm{T}}(\boldsymbol{y}(k) + \boldsymbol{A}\boldsymbol{p}(k) - \boldsymbol{b}))) \\ \boldsymbol{y}(k+1) = \boldsymbol{y}(k) + \boldsymbol{A}\boldsymbol{p}(k+1) - \boldsymbol{b} \end{cases} \quad (3.15)$$

定理 3.1 对于任意给定的初始值 $\boldsymbol{p}(0)$ 和 $\boldsymbol{y}(0)$，如果参数 σ 满足

$$\sigma < \frac{1}{2 + \lambda_{\max}(\boldsymbol{A}^{\mathrm{T}}\boldsymbol{A})}$$

那么算法(3.15)中的 $\boldsymbol{p}(k)$ 能够全局收敛到问题(3.13)的最优解，其中 $\lambda_{\max}$ 表示矩阵的最大特征值.

注 同样地，定理3.1的证明类似定理2.1的证明，将其留作练习.

3.5 分布式优化算法

上述对多机器人编队问题的处理采用的是集中式优化问题以及集中式优化算法，那么，分布式方法如何解决此类问题呢？

回顾前述的队形约束条件(3.6)，其表达式为

$$\boldsymbol{M}(\boldsymbol{p}_i - \boldsymbol{p}_1) = \boldsymbol{M}_i(\boldsymbol{p}_2 - \boldsymbol{p}_1) \quad (i = 3, 4, \cdots, m) \quad (3.16)$$

通过观察此公式，可以看出，机器人 i 需要知道如下的一些信息：自身的位置 $\boldsymbol{p}_i$，第 1 和第 2 号机器人的位置 $\boldsymbol{p}_1$ 和 $\boldsymbol{p}_2$，队形 $\boldsymbol{S}$ 中第 i $(i = 3, 4, \cdots, m)$ 个节点的位置生成的矩阵 $\boldsymbol{M}_i = \begin{pmatrix} s_i^x & -s_i^y \\ s_i^y & s_i^x \end{pmatrix}$，以及队形 $\boldsymbol{S}$ 中的第 2 个节点的位置生成的矩阵 $\boldsymbol{M} = \begin{pmatrix} \|\boldsymbol{s}_2\| & 0 \\ 0 & \|\boldsymbol{s}_2\| \end{pmatrix}$. 其中 $\boldsymbol{M}$ 和 $\boldsymbol{M}_i$ 是常量矩阵，其信息仅由第 i $(i = 3, 4, \cdots, m)$ 号机器人获取，即第 i 号机器人只需要知道队形 $\boldsymbol{S}$ 中的第 i 个节点的信息 $\boldsymbol{s}_i$ 和第 2 个节点的信息 $\boldsymbol{s}_2$. 然而，第 1 和第 2 号机器人的位置信息 $\boldsymbol{p}_1$ 和 $\boldsymbol{p}_2$ 需要在所有的机器人间进行共享，这样会增加这两个机器人的通信负载. 如图3.3所示，给出了 6 个机器人编队的例子，其中机器人 1 和机器人 2 需要与其他四个机器人进行通信，因此这两个机器

人的通信负载较大.

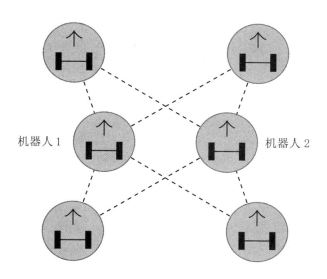

图 3.3 六个机器人形成的通信拓扑结构
注:机器人 1 和机器人 2 需要与其他四个机器人进行通信

接下来, 采用分布式的方法处理队形约束条件(3.16). 假设机器人 i ($i = 3, 4, \cdots, m$) 对机器人 1 和 2 的位置信息的估计量分别为 $\boldsymbol{p}_{i1}$ 和 $\boldsymbol{p}_{i2}$. 那么, 队形约束条件(3.16)可以转化为

$$\boldsymbol{M}(\boldsymbol{p}_i - \boldsymbol{p}_{i1}) = \boldsymbol{M}_i(\boldsymbol{p}_{i2} - \boldsymbol{p}_{i1}) \quad (i = 3, 4, \cdots, m) \tag{3.17}$$

然而, 原先的队形约束要求所有的机器人拥有相同的信息 $\boldsymbol{p}_1$ 和 $\boldsymbol{p}_2$, 因此, 需要添加如下的一致性约束条件:

$$\begin{cases} \boldsymbol{p}_{31} = \boldsymbol{p}_{41} = \cdots = \boldsymbol{p}_{m1} = \boldsymbol{p}_1 \\ \boldsymbol{p}_{32} = \boldsymbol{p}_{42} = \cdots = \boldsymbol{p}_{m2} = \boldsymbol{p}_2 \end{cases} \tag{3.18}$$

为了统一利用网络的整体拓扑结构进行分布式算法的设计, 加入机器人 1 对 $\boldsymbol{p}_2$ 的估计, 以及机器人 2 对 $\boldsymbol{p}_1$ 的估计. 因此, 一致性约束条件(3.18)可以进一步表示为

$$\begin{cases} \boldsymbol{p}_1 = \boldsymbol{p}_{21} = \boldsymbol{p}_{31} = \cdots = \boldsymbol{p}_{m1} \\ \boldsymbol{p}_{12} = \boldsymbol{p}_2 = \boldsymbol{p}_{32} = \cdots = \boldsymbol{p}_{m2} \end{cases} \tag{3.19}$$

假设 m 个机器人构成的通信网络图是连通的, 根据引理2.1, 等式约束条件(3.19)等价于

$$\begin{cases} \boldsymbol{L} \otimes \boldsymbol{I}_2 \boldsymbol{p}^1 = 0 \\ \boldsymbol{L} \otimes \boldsymbol{I}_2 \boldsymbol{p}^2 = 0 \end{cases} \tag{3.20}$$

其中 $\boldsymbol{L}$ 是 m 个机器人构成的通信网络图的拉普拉斯矩阵, $\boldsymbol{p}^1 = (\boldsymbol{p}_1^{\mathrm{T}}, \boldsymbol{p}_{21}^{\mathrm{T}}, \boldsymbol{p}_{31}^{\mathrm{T}}, \cdots, \boldsymbol{p}_{m1}^{\mathrm{T}})^{\mathrm{T}}$, $\boldsymbol{p}^2 = (\boldsymbol{p}_{12}^{\mathrm{T}}, \boldsymbol{p}_2^{\mathrm{T}}, \boldsymbol{p}_{32}^{\mathrm{T}}, \cdots, \boldsymbol{p}_{m2}^{\mathrm{T}})^{\mathrm{T}}$.

此外, 等式约束(3.17)可以写成

$$\begin{pmatrix} M & M_i - M & -M_i \end{pmatrix} \begin{pmatrix} p_i \\ p_{i1} \\ p_{i2} \end{pmatrix} = 0 \quad (i = 3, 4, \cdots, m)$$

设 $B_i = (M \ \ M_i - M \ \ -M_i)$, $\tilde{p}_i = (p_i^{\mathrm{T}}, p_{i1}^{\mathrm{T}}, p_{i2}^{\mathrm{T}})^{\mathrm{T}}$, 则上述等式约束可以表示为

$$B_i \tilde{p}_i = 0 \quad (i = 3, 4, \cdots, m) \tag{3.21}$$

如果加入锚点约束条件(3.10), 即 $p_1 - p_2 = b_3$, 用于约束机器人 1 和机器人 2 之间的相对位置, 也可以防止机器人在队形切换时出现间距过小的问题. 根据上面引入的记号, 可以对机器人 1 和机器人 2 分别加入约束条件 $p_1 - p_{12} = b_3$ 和 $p_2 - p_{21} = -b_3$.

综合上述的目标函数(3.12)、一致性约束(3.20)、队形约束(3.21)和锚点约束条件, 以及边界约束 $p_i \in \Omega_i$, 多机器人编队的分布式优化问题可以概括为

$$\min_{p_i} \quad \frac{1}{2} \sum_{i=1}^{m} \|p_i - \xi_i\|^2$$
$$\text{s.t.} \begin{cases} p_1 - p_{12} = b_3 \\ p_2 - p_{21} = -b_3 \\ B_i \tilde{p}_i = 0 \quad (i = 3, 4, \cdots, m) \\ L \otimes I_2 p^1 = 0 \\ L \otimes I_2 p^2 = 0 \\ p_i \in \Omega_i \quad (i = 1, 2, \cdots, m) \end{cases} \tag{3.22}$$

为了后续分析方便, 引入记号

$$\tilde{p} = (p_1^{\mathrm{T}}, p_{12}^{\mathrm{T}}, p_2^{\mathrm{T}}, p_{21}^{\mathrm{T}}, p_3^{\mathrm{T}}, p_{31}^{\mathrm{T}}, p_{32}^{\mathrm{T}}, \cdots, p_m^{\mathrm{T}}, p_{m1}^{\mathrm{T}}, p_{m2}^{\mathrm{T}})^{\mathrm{T}}$$
$$\tilde{c} = \mathrm{diag}\{1, 0, 1, 0, 1, 0, 0, \cdots, 1, 0, 0\}$$
$$\tilde{\xi} = (\xi_1^{\mathrm{T}}, 0_2^{\mathrm{T}}, \xi_2^{\mathrm{T}}, 0_2^{\mathrm{T}}, \xi_3^{\mathrm{T}}, 0_2^{\mathrm{T}}, 0_2^{\mathrm{T}}, \cdots, \xi_m^{\mathrm{T}}, 0_2^{\mathrm{T}}, 0_2^{\mathrm{T}})^{\mathrm{T}}$$

$$\tilde{B} = \begin{pmatrix} I_2 & -I_2 & O_2 & O_2 & O_{2 \times 6(m-2)} \\ O_2 & O_2 & I_2 & -I_2 & O_{2 \times 6(m-2)} \\ O_{2(m-2) \times 2} & O_{2(m-2) \times 2} & O_{2(m-2) \times 2} & O_{2(m-2) \times 2} & \mathrm{blkdiag}\{B_3, \cdots, B_m\} \end{pmatrix}$$

$$\tilde{b} = (b_3^{\mathrm{T}}, -b_3^{\mathrm{T}}, 0_{2(m-2)}^{\mathrm{T}})^{\mathrm{T}}$$

$$\tilde{\boldsymbol{L}}_1 = \begin{pmatrix} l_{11} & 0 & 0 & l_{12} & 0 & l_{13} & 0 & \cdots & 0 & l_{1m} & 0 \\ 0 & 0 & 0 & 0 & 0 & 0 & 0 & \cdots & 0 & 0 & 0 \\ 0 & 0 & 0 & 0 & 0 & 0 & 0 & \cdots & 0 & 0 & 0 \\ l_{21} & 0 & 0 & l_{22} & 0 & l_{23} & 0 & \cdots & 0 & l_{2m} & 0 \\ 0 & 0 & 0 & 0 & 0 & 0 & 0 & \cdots & 0 & 0 & 0 \\ l_{31} & 0 & 0 & l_{32} & 0 & l_{33} & 0 & \cdots & 0 & l_{3m} & 0 \\ 0 & 0 & 0 & 0 & 0 & 0 & 0 & \cdots & 0 & 0 & 0 \\ \vdots & \vdots & \vdots & \vdots & \vdots & \vdots & \vdots & & \vdots & \vdots & \vdots \\ 0 & 0 & 0 & 0 & 0 & 0 & 0 & \cdots & 0 & 0 & 0 \\ l_{m1} & 0 & 0 & l_{m2} & 0 & l_{m3} & 0 & \cdots & 0 & l_{mm} & 0 \\ 0 & 0 & 0 & 0 & 0 & 0 & 0 & \cdots & 0 & 0 & 0 \end{pmatrix}$$

$$\tilde{\boldsymbol{L}}_2 = \begin{pmatrix} 0 & 0 & 0 & 0 & 0 & 0 & 0 & \cdots & 0 & 0 & 0 \\ 0 & l_{11} & l_{12} & 0 & 0 & 0 & l_{13} & \cdots & 0 & 0 & l_{1m} \\ 0 & l_{21} & l_{22} & 0 & 0 & 0 & l_{23} & \cdots & 0 & 0 & l_{2m} \\ 0 & 0 & 0 & 0 & 0 & 0 & 0 & \cdots & 0 & 0 & 0 \\ 0 & 0 & 0 & 0 & 0 & 0 & 0 & \cdots & 0 & 0 & 0 \\ 0 & 0 & 0 & 0 & 0 & 0 & 0 & \cdots & 0 & 0 & 0 \\ 0 & l_{31} & l_{32} & 0 & 0 & 0 & l_{33} & \cdots & 0 & 0 & l_{3m} \\ \vdots & \vdots & \vdots & \vdots & \vdots & \vdots & \vdots & & \vdots & \vdots & \vdots \\ 0 & 0 & 0 & 0 & 0 & 0 & 0 & \cdots & 0 & 0 & 0 \\ 0 & 0 & 0 & 0 & 0 & 0 & 0 & \cdots & 0 & 0 & 0 \\ 0 & l_{m1} & l_{m2} & 0 & 0 & 0 & l_{m3} & \cdots & 0 & 0 & l_{mm} \end{pmatrix}$$

$$\tilde{\Omega} = \Omega_1 \times \Omega_2 \times \Omega_2 \times \Omega_1 \times \Omega_3 \times \Omega_1 \times \Omega_2 \times \cdots \times \Omega_m \times \Omega_1 \times \Omega_2$$

其中 diag{·} 表示对角矩阵, blkdiag{·} 表示分块对角矩阵.

因此, 优化问题(3.22)可以简写成如下的形式:

$$\min_{\boldsymbol{p}} \quad \frac{1}{2}\|\tilde{\boldsymbol{c}}\tilde{\boldsymbol{p}} - \tilde{\boldsymbol{\xi}}\|^2$$
$$\text{s.t.} \begin{cases} \tilde{\boldsymbol{B}}\tilde{\boldsymbol{p}} = \tilde{\boldsymbol{b}} \\ (\tilde{\boldsymbol{L}}_1 \otimes \boldsymbol{I}_2)\tilde{\boldsymbol{p}} = 0 \\ (\tilde{\boldsymbol{L}}_2 \otimes \boldsymbol{I}_2)\tilde{\boldsymbol{p}} = 0 \\ \tilde{\boldsymbol{p}} \in \tilde{\Omega} \end{cases} \quad (3.23)$$

对于优化问题(3.23)的求解,可以采用和前述类似的方法建立如下的算法公式:

$$\begin{cases} \tilde{\boldsymbol{p}}(k+1) &= \phi(\tilde{\boldsymbol{p}}(k) - \sigma(\tilde{\boldsymbol{c}}\tilde{\boldsymbol{p}}(k) - \tilde{\boldsymbol{\xi}} + \tilde{\boldsymbol{B}}^{\mathrm{T}}(\tilde{\boldsymbol{y}}(k) + \tilde{\boldsymbol{B}}\tilde{\boldsymbol{p}}(k) - \tilde{\boldsymbol{b}}) \\ & \quad + \tilde{\boldsymbol{z}} + (\tilde{\boldsymbol{L}}_1 \otimes \boldsymbol{I}_2)\tilde{\boldsymbol{p}} + \tilde{\boldsymbol{w}} + (\tilde{\boldsymbol{L}}_2 \otimes \boldsymbol{I}_2)\tilde{\boldsymbol{p}})) \\ \tilde{\boldsymbol{y}}(k+1) &= \tilde{\boldsymbol{y}}(k) + \tilde{\boldsymbol{B}}\tilde{\boldsymbol{p}}(k+1) - \tilde{\boldsymbol{b}} \\ \tilde{\boldsymbol{z}}(k+1) &= \tilde{\boldsymbol{z}}(k) + (\tilde{\boldsymbol{L}}_1 \otimes \boldsymbol{I}_2)\tilde{\boldsymbol{p}}(k+1) \\ \tilde{\boldsymbol{w}}(k+1) &= \tilde{\boldsymbol{w}}(k) + (\tilde{\boldsymbol{L}}_2 \otimes \boldsymbol{I}_2)\tilde{\boldsymbol{p}}(k+1) \end{cases} \quad (3.24)$$

其中 $\tilde{\boldsymbol{z}} = (\boldsymbol{z}_1^{\mathrm{T}}, \boldsymbol{z}_{12}^{\mathrm{T}}, \boldsymbol{z}_2^{\mathrm{T}}, \boldsymbol{z}_{21}^{\mathrm{T}}, \boldsymbol{z}_3^{\mathrm{T}}, \boldsymbol{z}_{31}^{\mathrm{T}}, \boldsymbol{z}_{32}^{\mathrm{T}}, \cdots, \boldsymbol{z}_m^{\mathrm{T}}, \boldsymbol{z}_{m1}^{\mathrm{T}}, \boldsymbol{z}_{m2}^{\mathrm{T}})^{\mathrm{T}}$, $\tilde{\boldsymbol{w}} = (\boldsymbol{w}_1^{\mathrm{T}}, \boldsymbol{w}_{12}^{\mathrm{T}}, \boldsymbol{w}_2^{\mathrm{T}}, \boldsymbol{w}_{21}^{\mathrm{T}}, \boldsymbol{w}_3^{\mathrm{T}}, \boldsymbol{w}_{31}^{\mathrm{T}}, \boldsymbol{w}_{32}^{\mathrm{T}}, \cdots, \boldsymbol{w}_m^{\mathrm{T}}, \boldsymbol{w}_{m1}^{\mathrm{T}}, \boldsymbol{w}_{m2}^{\mathrm{T}})^{\mathrm{T}}$, 采用和 $\tilde{\boldsymbol{p}}$ 同样的下标以方便书写; $\tilde{\boldsymbol{y}} = (\boldsymbol{y}_1^{\mathrm{T}}, \boldsymbol{y}_2^{\mathrm{T}}, \boldsymbol{y}_3^{\mathrm{T}}, \cdots, \boldsymbol{y}_m^{\mathrm{T}})^{\mathrm{T}}$, 其中 $\boldsymbol{y}_i \in \mathbf{R}^2$ $(i = 1, 2, 3, \cdots, m)$.

接下来, 将(3.24)写成分量的形式, 可以清楚地看出算法的分布式处理方式.

对于机器人 1, 其迭代公式为

$$\begin{cases} \boldsymbol{p}_1(k+1) &= \phi_1(\boldsymbol{p}_1(k) - \sigma(\boldsymbol{p}_1(k) - \boldsymbol{\xi}_1 + \boldsymbol{y}_1(k) + \boldsymbol{p}_1(k) - \boldsymbol{p}_{12}(k) - \boldsymbol{b}_3 \\ & \quad + \boldsymbol{z}_1(k) + \sum_{j \in N_1} a_{1j}(\boldsymbol{p}_1(k) - \boldsymbol{p}_{j1}(k)))) \\ \boldsymbol{p}_{12}(k+1) &= \phi_2(\boldsymbol{p}_{12}(k) - \sigma(-\boldsymbol{y}_1(k) - \boldsymbol{p}_1(k) + \boldsymbol{p}_{12}(k) + \boldsymbol{b}_3 \\ & \quad + \boldsymbol{w}_{12}(k) + \sum_{j \in N_1} a_{1j}(\boldsymbol{p}_{12}(k) - \boldsymbol{p}_{j2}(k)))) \\ \boldsymbol{y}_1(k+1) &= \boldsymbol{y}_1(k) + \boldsymbol{p}_1(k+1) - \boldsymbol{p}_{12}(k+1) - \boldsymbol{b}_3 \\ \boldsymbol{z}_1(k+1) &= \boldsymbol{z}_1(k) + \sum_{j \in N_1} a_{1j}(\boldsymbol{p}_1(k+1) - \boldsymbol{p}_{j1}(k+1)) \\ \boldsymbol{w}_{12}(k+1) &= \boldsymbol{w}_{12}(k) + \sum_{j \in N_1} a_{1j}(\boldsymbol{p}_{12}(k+1) - \boldsymbol{p}_{j2}(k+1)) \end{cases} \quad (3.25)$$

其中当 $a_{12} \neq 0$ 时, $\boldsymbol{p}_{22} = \boldsymbol{p}_2$.

对于机器人 2, 其迭代公式为

$$\begin{cases} \boldsymbol{p}_2(k+1) &= \phi_2(\boldsymbol{p}_2(k) - \sigma(\boldsymbol{p}_2(k) - \boldsymbol{\xi}_2 + \boldsymbol{y}_2(k) + \boldsymbol{p}_2(k) - \boldsymbol{p}_{21}(k) + \boldsymbol{b}_3 \\ & \quad + \boldsymbol{w}_2(k) + \sum_{j \in N_2} a_{2j}(\boldsymbol{p}_2(k) - \boldsymbol{p}_{j2}(k)))) \\ \boldsymbol{p}_{21}(k+1) &= \phi_1(\boldsymbol{p}_{21}(k) - \sigma(-\boldsymbol{y}_2(k) - \boldsymbol{p}_2(k) + \boldsymbol{p}_{21}(k) - \boldsymbol{b}_3 \\ & \quad + \boldsymbol{z}_{21}(k) + \sum_{j \in N_2} a_{2j}(\boldsymbol{p}_{21}(k) - \boldsymbol{p}_{j1}(k)))) \\ \boldsymbol{y}_2(k+1) &= \boldsymbol{y}_2(k) + \boldsymbol{p}_2(k+1) - \boldsymbol{p}_{21}(k+1) + \boldsymbol{b}_3 \\ \boldsymbol{z}_{21}(k+1) &= \boldsymbol{z}_{21}(k) + \sum_{j \in N_2} a_{2j}(\boldsymbol{p}_{21}(k+1) - \boldsymbol{p}_{j1}(k+1)) \\ \boldsymbol{w}_2(k+1) &= \boldsymbol{w}_2(k) + \sum_{j \in N_2} a_{2j}(\boldsymbol{p}_2(k+1) - \boldsymbol{p}_{j2}(k+1)) \end{cases} \quad (3.26)$$

其中当 $a_{21} \neq 0$ 时, $\boldsymbol{p}_{11} = \boldsymbol{p}_1$.

对于机器人 i $(i=3,4,\cdots,m)$，其迭代公式为

$$\begin{cases} \boldsymbol{p}_i(k+1) &= \phi_i(\boldsymbol{p}_i(k) - \sigma(\boldsymbol{p}_i(k) - \boldsymbol{\xi}_i + \boldsymbol{M}(\boldsymbol{y}_i(k) + \boldsymbol{B}_i\tilde{\boldsymbol{p}}_i(k)))) \\ \boldsymbol{p}_{i1}(k+1) &= \phi_1(\boldsymbol{p}_{i1}(k) - \sigma((\boldsymbol{M}_i^{\mathrm{T}} - \boldsymbol{M})(\boldsymbol{y}_i(k) + \boldsymbol{B}_i\tilde{\boldsymbol{p}}_i(k)) + \boldsymbol{z}_{i1}(k) \\ & \quad + \sum_{j \in N_i} a_{ij}(\boldsymbol{p}_{i1}(k) - \boldsymbol{p}_{j1}(k)))) \\ \boldsymbol{p}_{i2}(k+1) &= \phi_2(\boldsymbol{p}_{i2}(k) - \sigma(-\boldsymbol{M}_i^{\mathrm{T}}(\boldsymbol{y}_i(k) + \boldsymbol{B}_i\tilde{\boldsymbol{p}}_i(k)) + \boldsymbol{w}_{i2}(k) \\ & \quad + \sum_{j \in N_i} a_{ij}(\boldsymbol{p}_{i2}(k) - \boldsymbol{p}_{j2}(k)))) \\ \boldsymbol{y}_i(k+1) &= \boldsymbol{y}_i(k) + \boldsymbol{B}_i\tilde{\boldsymbol{p}}_i(k+1) \\ \boldsymbol{z}_{i1}(k+1) &= \boldsymbol{z}_{i1}(k) + \sum_{j \in N_i} a_{ij}(\boldsymbol{p}_{i1}(k+1) - \boldsymbol{p}_{j1}(k+1)) \\ \boldsymbol{w}_{i2}(k+1) &= \boldsymbol{w}_{i2}(k) + \sum_{j \in N_i} a_{ij}(\boldsymbol{p}_{i2}(k+1) - \boldsymbol{p}_{j2}(k+1)) \end{cases} \quad (3.27)$$

其中当 $a_{i1} \neq 0$ 时, $\boldsymbol{p}_{11} = \boldsymbol{p}_1$; 当 $a_{i2} \neq 0$ 时, $\boldsymbol{p}_{22} = \boldsymbol{p}_2$; $\tilde{\boldsymbol{p}}_i = (\boldsymbol{p}_i^{\mathrm{T}}, \boldsymbol{p}_{i1}^{\mathrm{T}}, \boldsymbol{p}_{i2}^{\mathrm{T}})^{\mathrm{T}}$.

当采用迭代公式(3.25)~(3.27)对优化问题(3.23)进行求解时, 只需要机器人群构成的通信网络图是连通的即可. 如图3.4所示, 给出了 6 个机器人编队的例子, 其中 6 个机器人构成了一个环形的通信拓扑图, 机器人 1 和机器人 2 只需要分别与其他两个机器人进行通信, 因此可以极大地降低机器人的通信负载.

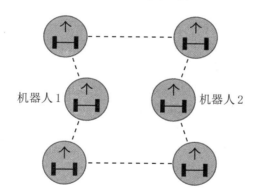

图 3.4 六个机器人形成的通信拓扑结构
注: 机器人 1 和机器人 2 需要与其他四个机器人进行通信

❦ 练 习 ❧

1. 证明引理3.1.
2. 证明定理3.1.
3. 假设有 6 个机器人进行编队, 对于分布式优化算法(3.25)~(3.27), 对于图3.4的环形通信拓扑结构写出每个机器人对应的迭代算法.

第 4 章 时变分布式优化与多机器人实时围堵

前两章介绍的分布式优化问题都是静态的,即优化问题的目标函数和约束条件都与时间无关. 本章将介绍时变分布式优化问题,并研究其在解决多机器人协同围堵问题中的应用. 首先将建立实现这类多机器人围堵问题的时变优化模型,进而设计解决此类问题的时变分布式优化算法.

4.1 多机器人实时围堵问题

首先建立实现多机器人实时围堵任务的优化模型. 假设有 m 个机器人要完成对 0 号机器人的围堵任务. 已知 0 号机器人的运动轨迹是与时间 t 相关的曲线,记为 $c(t)$. m 个机器人以一定的队形实现对 0 号机器人的围堵,但同时需要实现如下三个任务:

(1) 围堵机器人有各自的监测点,期望围堵机器人到监测点的总移动距离和最小;
(2) 为了能够实现有效围堵,期望围堵机器人到围堵目标的总移动距离和最小;
(3) 围堵目标处于围堵机器人群的几何中心.

图4.1给出了实现机器人围堵任务的一个示意图. 图中机器人 0 是被围堵对象,另外 6 个机器人实现对机器人 0 的围堵,同时 6 个围堵机器人通过环形拓扑结构进行信息通信.

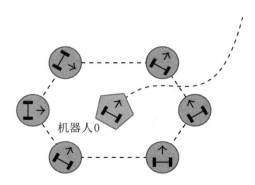

图 4.1　六个机器人形成对机器人 0 的围堵

假设有 m 个机器人来完成围堵任务. 围堵目标 G 的轨迹为 $c(t)$;执行围堵任务的机器人 i 的空间坐标为 p_i;机器人 i 的监测点坐标为 $s_i(t)$, s_i 对目标 G 形成合围的态势.

为了实现上述的第一个任务, 即围堵机器人到监测点的总移动距离和最小, 需要最小化如下的目标函数:

$$\sum_{i=1}^{m}\|\boldsymbol{p}_i - \boldsymbol{s}_i(t)\| \tag{4.1}$$

为了实现上述的第二个任务, 即围堵机器人到围堵目标的总移动距离和最小, 需要最小化如下的目标函数:

$$\sum_{i=1}^{m}\|\boldsymbol{p}_i - \boldsymbol{c}(t)\| \tag{4.2}$$

为了实现上述的第三个任务, 即围堵目标处于围堵机器人群的几何中心, 围堵机器人和围堵目标需要满足如下的等式约束条件:

$$\frac{1}{m}\sum_{i=1}^{m}\boldsymbol{p}_i = \boldsymbol{c}(t) \tag{4.3}$$

综合(4.1)~(4.3), 实现多机器人围堵任务的时变优化问题可以表示为

$$\begin{aligned}\min_{\boldsymbol{p}_i} \quad & \frac{1}{2}\sum_{i=1}^{m}\left(\alpha_i\|\boldsymbol{p}_i - \boldsymbol{s}_i(t)\|^2 + (1-\alpha_i)\|\boldsymbol{p}_i - \boldsymbol{c}(t)\|^2\right) \\ \text{s.t.} \quad & \sum_{i=1}^{m}\boldsymbol{p}_i = m\boldsymbol{c}(t)\end{aligned} \tag{4.4}$$

其中 $0 \leqslant \alpha_i \leqslant 1$ 是权重系数, 用以调节围堵机器人到围堵目标的距离与围堵机器人到监测点的距离的比例. 当 α 接近 0 时, 表示围堵机器人更想要靠近监测点; 当 α 接近 1 时, 表示围堵机器人更想要靠近围堵目标.

注 优化问题(4.4)中的已知量 $\boldsymbol{c}(t)$ 和 $\boldsymbol{s}_i(t)$ 都是随时间变化的函数, 此类优化问题称为**时变优化**问题. 时变优化问题不同于前面两章介绍的时不变优化问题, 其算法设计和理论分析也较困难.

多机器人系统可以看作多智能体系统 (multi-agent system) 的一种具体形式, 关于多智能体系统的一致性理论可以具体应用到多机器人系统. 接下来, 将介绍一个关于多智能体系统的一致性算法, 它将用于后续的理论分析.

首先, 考虑一个由 m 个个体组成的多智能体系统, 其第 i 个智能体的状态由如下方程描述:

$$\dot{\boldsymbol{z}}_i(t) = -\boldsymbol{\Xi}\boldsymbol{z}_i(t) + \boldsymbol{v}_i(t) - \alpha\sum_{j\in N_i}\text{sign}(\boldsymbol{z}_i(t) - \boldsymbol{z}_j(t)) \tag{4.5}$$

其中 $\boldsymbol{z}_i(t) \in \mathbf{R}^n$ 是智能体 i 的状态向量, $\boldsymbol{\Xi}$ 是一个常系数矩阵, α 是正常数, $\text{sign}(\cdot)$ 是符号函数, N_i 是智能体 i 的邻居集合, $\boldsymbol{v}_i(t) \in \mathbf{R}^n$ 是有界的时变向量函数, 满足

$$\sup_{t\geqslant 0}\|\boldsymbol{v}_i(t)\|_\infty < \kappa.$$

引理 4.1 (渐近一致性)

假设多智能体系统(4.5)的通信网络图是无向连通的, 如果参数 α 满足 $\alpha \geqslant \kappa\sqrt{2nm/\lambda_2(\boldsymbol{L})}$, 其中 $\lambda_2(\boldsymbol{L})$ 是矩阵 $\boldsymbol{L}$ 的第二最小特征值, 且系数矩阵 $\boldsymbol{\Xi}$ 满足 $\boldsymbol{\Xi} + \boldsymbol{\Xi}^{\mathrm{T}}$ 是正定的, 则多智能体系统的状态向量 $\boldsymbol{z}_i(t)$ $(i=1,2,\cdots,m)$ 当 $t \to \infty$ 时达到渐近一致.

证明 记 $\bar{\boldsymbol{z}}(t) = (1/n)\sum_{j=1}^{m}\boldsymbol{z}_j(t)$, $\boldsymbol{e}_i(t) = \boldsymbol{z}_i(t) - \bar{\boldsymbol{z}}(t)$. 因为多智能体系统的通信网络图是无向的, 有

$$\dot{\bar{\boldsymbol{z}}}(t) = \frac{1}{m}\sum_{j=1}^{m}\dot{\boldsymbol{z}}_j(t) = \frac{1}{m}\sum_{j=1}^{m}\big(-\boldsymbol{\Xi}\boldsymbol{z}_j(t) + \boldsymbol{v}_j(t)\big)$$

进而得出

$$\begin{aligned}\dot{\boldsymbol{z}}_i(t) - \dot{\bar{\boldsymbol{z}}}(t) = &-\boldsymbol{\Xi}\boldsymbol{z}_i(t) + \frac{1}{m}\sum_{j=1}^{m}\boldsymbol{\Xi}\boldsymbol{z}_j(t) + \boldsymbol{v}_i(t) \\ &- \frac{1}{m}\sum_{j=1}^{m}\boldsymbol{v}_j(t) - \alpha\sum_{j\in N_i}\mathrm{sign}(\boldsymbol{z}_i(t) - \boldsymbol{z}_j(t))\end{aligned} \quad (4.6)$$

引入李雅普诺夫函数 $V(t) = (1/2)\sum_{i=1}^{m}\boldsymbol{e}_i^{\mathrm{T}}(t)\boldsymbol{e}_i(t)$, 可以计算 $V(t)$ 在系统(4.6)下的导数为

$$\begin{aligned}\dot{V}(t) &= \sum_{i=1}^{m}\boldsymbol{e}_i^{\mathrm{T}}(t)\dot{\boldsymbol{e}}_i(t) = \sum_{i=1}^{m}\boldsymbol{e}_i^{\mathrm{T}}(t)\big(\dot{\boldsymbol{z}}_i(t) - \dot{\bar{\boldsymbol{z}}}(t)\big) \\ &= \sum_{i=1}^{m}\boldsymbol{e}_i^{\mathrm{T}}(t)\Big(-\boldsymbol{\Xi}\big(\boldsymbol{z}_i(t) - \frac{1}{m}\sum_{j=1}^{m}\boldsymbol{z}_j(t)\big) \\ &\quad + \boldsymbol{v}_i(t) - \frac{1}{m}\sum_{j=1}^{m}\boldsymbol{v}_j(t) - \alpha\sum_{j\in N_i}\mathrm{sign}(\boldsymbol{z}_i(t) - \boldsymbol{z}_j(t))\Big) \\ &= \sum_{i=1}^{m}\boldsymbol{e}_i^{\mathrm{T}}(t)\Big(-\boldsymbol{\Xi}\boldsymbol{e}_i(t) + \boldsymbol{v}_i(t) - \alpha\sum_{j\in N_i}\mathrm{sign}(\boldsymbol{z}_i(t) - \boldsymbol{z}_j(t))\Big)\end{aligned} \quad (4.7)$$

其中最后一个等式成立是因为 $\boldsymbol{e}_i = \boldsymbol{z}_i(t) - (1/n)\sum_{j=1}^{m}\boldsymbol{z}_j(t)$ 和 $\sum_{i=1}^{m}\boldsymbol{e}_i^{\mathrm{T}}(t)\sum_{j=1}^{m}\boldsymbol{v}_j(t) = 0$.

由于多智能体系统的通信网络图是无向的, 有

$$\sum_{i=1}^{m} \boldsymbol{e}_i^{\mathrm{T}}(t) \sum_{j \in N_i} \mathrm{sign}(\boldsymbol{z}_i(t) - \boldsymbol{z}_j(t))$$
$$= \frac{1}{2} \sum_{i=1}^{m} \sum_{j \in N_i} (\boldsymbol{e}_i(t) - \boldsymbol{e}_j(t))^{\mathrm{T}} \mathrm{sign}(\boldsymbol{z}_i(t) - \boldsymbol{z}_j(t))$$
$$= \frac{1}{2} \sum_{i=1}^{m} \sum_{j \in N_i} (\boldsymbol{e}_i(t) - \boldsymbol{e}_j(t))^{\mathrm{T}} \mathrm{sign}(\boldsymbol{e}_i(t) - \boldsymbol{e}_j(t))$$
$$= \frac{1}{2} \sum_{i=1}^{m} \sum_{j \in N_i} \|\boldsymbol{e}_i(t) - \boldsymbol{e}_j(t)\|_1 \tag{4.8}$$

其中第一个等式是根据文献 [12] 的引理 6.1 推导得到的, 第二个等式根据 $\boldsymbol{e}_i(t)$ 的定义可以得到.

将(4.8)式代入(4.7)式可以推出

$$\dot{V}(t) = -\sum_{i=1}^{m} \boldsymbol{e}_i^{\mathrm{T}}(t) \boldsymbol{\Xi} \boldsymbol{e}_i(t) + \sum_{i=1}^{m} \boldsymbol{e}_i^{\mathrm{T}}(t) \boldsymbol{v}_i(t) - \frac{\alpha}{2} \sum_{i=1}^{m} \sum_{j \in N_i} \|\boldsymbol{e}_i(t) - \boldsymbol{e}_j(t)\|_1$$
$$\leqslant -\frac{1}{2} \sum_{i=1}^{m} \boldsymbol{e}_i^{\mathrm{T}}(t)(\boldsymbol{\Xi} + \boldsymbol{\Xi}^{\mathrm{T}}) \boldsymbol{e}_i(t) + \sum_{i=1}^{m} \|\boldsymbol{e}_i(t)\|_1 \|\boldsymbol{v}_i(t)\|_\infty - \frac{\alpha}{2} \sum_{i=1}^{m} \sum_{j \in N_i} \|\boldsymbol{e}_i(t) - \boldsymbol{e}_j(t)\|_1$$
$$\leqslant -\tau \sum_{i=1}^{m} \|\boldsymbol{e}_i(t)\|^2 + \kappa \sum_{i=1}^{m} \|\boldsymbol{e}_i(t)\|_1 - \frac{\alpha}{2} \sum_{i=1}^{m} \sum_{j \in N_i} \|\boldsymbol{e}_i(t) - \boldsymbol{e}_j(t)\|_1 \tag{4.9}$$

其中第二个不等式成立是因为 $\sup\limits_{t \geqslant 0} \|\boldsymbol{v}_i(t)\|_\infty < \kappa$, 以及矩阵 $\boldsymbol{\Xi} + \boldsymbol{\Xi}^{\mathrm{T}}$ 是正定的, 即存在 $\tau > 0$ 使得 $\boldsymbol{e}_i^{\mathrm{T}}(t)(\boldsymbol{\Xi} + \boldsymbol{\Xi}^{\mathrm{T}}) \boldsymbol{e}_i(t) \geqslant 2\tau \|\boldsymbol{e}_i(t)\|^2$.

根据范数的性质, 有 $\sum_{i=1}^{m} \|\boldsymbol{e}_i(t)\|_1 \leqslant \sqrt{nm} (\sum_{i=1}^{m} \|\boldsymbol{e}_i(t)\|^2)^{1/2}$ 和 $\|\boldsymbol{e}_i(t) - \boldsymbol{e}_j(t)\|_1 \geqslant \|\boldsymbol{e}_i(t) - \boldsymbol{e}_j(t)\|$. 那么(4.9)式可以推出

$$\dot{V}(t) \leqslant -\tau \sum_{i=1}^{m} \|\boldsymbol{e}_i(t)\|^2 + \kappa \sqrt{nm} \left(\sum_{i=1}^{m} \|\boldsymbol{e}_i(t)\|^2 \right)^{\frac{1}{2}} - \frac{\alpha}{2} \sum_{i=1}^{m} \sum_{j \in N_i} \|\boldsymbol{e}_i(t) - \boldsymbol{e}_j(t)\|$$
$$\leqslant -\tau \sum_{i=1}^{m} \|\boldsymbol{e}_i(t)\|^2 + \kappa \sqrt{nm} \left(\sum_{i=1}^{m} \|\boldsymbol{e}_i(t)\|^2 \right)^{\frac{1}{2}} - \frac{\alpha}{2} \left(\sum_{i=1}^{m} \sum_{j \in N_i} \|\boldsymbol{e}_i(t) - \boldsymbol{e}_j(t)\|^2 \right)^{\frac{1}{2}}$$
$$\leqslant -\tau \sum_{i=1}^{m} \|\boldsymbol{e}_i(t)\|^2 + \kappa \sqrt{nm} (2V(t))^{\frac{1}{2}} - \frac{\alpha}{2} \left(2\lambda_2(L) \sum_{i=1}^{m} \|\boldsymbol{e}_i(t)\|^2 \right)^{\frac{1}{2}}$$
$$= -2\tau V(t) - (\alpha \sqrt{\lambda_2(L)} - \kappa \sqrt{2nm}) V^{\frac{1}{2}}$$

其中第三个不等式成立是由于对于任意的 $\mathbf{1}_m^{\mathrm{T}}x = 0$ 且 $x \neq 0$, $\sum_{i=1}^{m}\sum_{j \in N_i}(x_i - x_j)^2 = 2x^{\mathrm{T}}\boldsymbol{L}x \geqslant 2\lambda_2(\boldsymbol{L})x^{\mathrm{T}}x$ 成立 [9].

如果 $\alpha \geqslant \kappa\sqrt{2nm/\lambda_2(\boldsymbol{L})}$ 成立, 那么 $V(t)$ 将指数收敛到零. 也就是说多智能体系统(4.5)的状态向量 $\boldsymbol{z}_i(t)$ $(i = 1, 2, \cdots, m)$ 当 $t \to \infty$ 时达到渐近一致. 证毕!

4.2 集中式优化算法

记 $f_i(\boldsymbol{p}_i, t) = (\eta_i\|\boldsymbol{p}_i - \boldsymbol{s}_i(t)\|^2 + (1 - \eta_i)\|\boldsymbol{p}_i - \boldsymbol{c}(t)\|^2)/2$, $\boldsymbol{b}(t) = m\boldsymbol{c}(t)$, 那么优化问题(4.4)可以写成

$$\begin{aligned}\min_{\boldsymbol{p}_i} \quad & \sum_{i=1}^{m} f_i(\boldsymbol{p}_i, t) \\ \text{s.t.} \quad & \sum_{i=1}^{m} \boldsymbol{p}_i = \boldsymbol{b}(t)\end{aligned} \quad (4.10)$$

优化问题(4.10)的拉格朗日函数定义为

$$\Psi(\boldsymbol{p}, \boldsymbol{\mu}, t) = \sum_{i=1}^{m} f_i(\boldsymbol{p}_i, t) + \boldsymbol{\mu}^{\mathrm{T}}(t)\left(\sum_{i=1}^{m}\boldsymbol{p}_i - \boldsymbol{b}(t)\right) \quad (4.11)$$

其中 $\boldsymbol{p} = (\boldsymbol{p}_1^{\mathrm{T}}, \boldsymbol{p}_2^{\mathrm{T}}, \cdots, \boldsymbol{p}_m^{\mathrm{T}})^{\mathrm{T}}$, $\boldsymbol{\mu}(t) \in \mathbf{R}^n$ 是拉格朗日乘子 (为了简化起见, 经常写成 $\boldsymbol{\mu}$). 如果分解 $\boldsymbol{b}(t)$ 使得 $\sum_{i=1}^{m}\boldsymbol{b}_i(t) = \boldsymbol{b}(t)$, 则拉格朗日函数可以重写为

$$\Psi(\boldsymbol{p}, \boldsymbol{\mu}, t) = \sum_{i=1}^{m} f_i(\boldsymbol{p}_i, t) + \boldsymbol{\mu}^{\mathrm{T}}\left(\sum_{i=1}^{m}(\boldsymbol{p}_i - \boldsymbol{b}_i(t))\right) \quad (4.12)$$

进而拉格朗日函数的海塞矩阵 (Hessian matrix) 可以表示为

$$\tilde{\boldsymbol{H}} = \begin{pmatrix} \boldsymbol{I} & \boldsymbol{O} & \cdots & \boldsymbol{O} & \boldsymbol{I} \\ \boldsymbol{O} & \boldsymbol{I} & \cdots & \boldsymbol{O} & \boldsymbol{I} \\ \vdots & \vdots & \ddots & \vdots & \vdots \\ \boldsymbol{O} & \boldsymbol{O} & \cdots & \boldsymbol{I} & \boldsymbol{I} \\ \boldsymbol{I} & \boldsymbol{I} & \cdots & \boldsymbol{I} & \boldsymbol{O} \end{pmatrix} \quad (4.13)$$

其中 $\boldsymbol{O}$ 是 n 维的零矩阵, $\boldsymbol{I}$ 是 n 维的单位矩阵.

基于预测修正方法 (prediction-correction method)[1,6], 如果向量 $(\boldsymbol{x}^{\mathrm{T}}, \boldsymbol{\mu}^{\mathrm{T}})^{\mathrm{T}}$ 满足

$$\tilde{\boldsymbol{H}}\begin{pmatrix}\dot{\boldsymbol{x}}(t)\\ \dot{\boldsymbol{\mu}}(t)\end{pmatrix} = -\sigma\begin{pmatrix}\boldsymbol{\nabla}_x\Psi(\boldsymbol{x},\boldsymbol{\mu},t)\\ \boldsymbol{\nabla}_\mu\Psi(\boldsymbol{x},\boldsymbol{\mu},t)\end{pmatrix} - \begin{pmatrix}\boldsymbol{\nabla}_{xt}\Psi(\boldsymbol{x},\boldsymbol{\mu},t)\\ \boldsymbol{\nabla}_{\mu t}\Psi(\boldsymbol{x},\boldsymbol{\mu},t)\end{pmatrix} \tag{4.14}$$

则对于 $\sigma > 0$, 可以得到 $(\boldsymbol{x}^{\mathrm{T}}, \boldsymbol{\mu}^{\mathrm{T}})^{\mathrm{T}}$ 全局指数收敛到最优解 $(\boldsymbol{x}^{*\mathrm{T}}(t), \boldsymbol{\mu}^{*\mathrm{T}}(t))^{\mathrm{T}}$.

结合式(4.11)和式(4.13), 第 i 个智能体的系统方程(4.14)可以重新写成

$$\begin{align}
\dot{\boldsymbol{p}}_i + \dot{\boldsymbol{\mu}} &= -\sigma\boldsymbol{\nabla}_{\boldsymbol{p}_i}f_i(\boldsymbol{p}_i,t) - \sigma\boldsymbol{\mu} - \boldsymbol{\nabla}_{\boldsymbol{p}_it}f_i(\boldsymbol{p}_i,t) \tag{4.15a}\\
\sum_{i=1}^{m}\dot{\boldsymbol{p}}_i &= -\sigma\left(\sum_{i=1}^{m}\boldsymbol{p}_i - \boldsymbol{b}(t)\right) + \dot{\boldsymbol{b}}(t) \tag{4.15b}
\end{align}$$

其中 $\boldsymbol{\nabla}_{\boldsymbol{p}_i}f_i(\boldsymbol{p}_i,t) = \boldsymbol{p}_i - \eta_i\boldsymbol{s}_i(t) - (1-\eta_i)\boldsymbol{c}(t)$, $\boldsymbol{\nabla}_{\boldsymbol{p}_it}f_i(\boldsymbol{p}_i,t) = -\eta_i\dot{\boldsymbol{s}}_i(t) - (1-\eta_i)\dot{\boldsymbol{c}}(t)$.

根据方程(4.15a), 得到

$$\dot{\boldsymbol{p}}_i = -\dot{\boldsymbol{\mu}} - \sigma\boldsymbol{\nabla}_{\boldsymbol{p}_i}f_i(\boldsymbol{p}_i,t) - \sigma\boldsymbol{\mu} - \boldsymbol{\nabla}_{\boldsymbol{p}_it}f_i(\boldsymbol{p}_i,t)$$

对上式两边从 1 到 m 进行求和, 得到

$$\sum_{i=1}^{m}\dot{\boldsymbol{p}}_i = -\sum_{i=1}^{m}\left(\dot{\boldsymbol{\mu}} + \sigma\boldsymbol{\nabla}_{\boldsymbol{p}_i}f_i(\boldsymbol{p}_i,t) + \sigma\boldsymbol{\mu} + \boldsymbol{\nabla}_{\boldsymbol{p}_it}f_i(\boldsymbol{p}_i,t)\right) \tag{4.16}$$

进而, 将(4.16)式代入(4.15b)式得到

$$\begin{align}
\dot{\boldsymbol{\mu}} &= \frac{1}{m}\left[-\sum_{i=1}^{m}\left(\sigma\boldsymbol{\nabla}_{\boldsymbol{p}_i}f_i(\boldsymbol{p}_i,t) + \sigma\boldsymbol{\mu} + \boldsymbol{\nabla}_{\boldsymbol{p}_it}f_i(\boldsymbol{p}_i,t)\right) + \sigma\left(\sum_{i=1}^{m}\boldsymbol{p}_i - \boldsymbol{b}(t)\right) - \dot{\boldsymbol{b}}(t)\right]\\
&= \frac{1}{m}\left[-\sum_{i=1}^{m}\left(\sigma\boldsymbol{\nabla}_{\boldsymbol{p}_i}f_i(\boldsymbol{p}_i,t) + \boldsymbol{\nabla}_{\boldsymbol{p}_it}f_i(\boldsymbol{p}_i,t)\right) + \sigma\left(\sum_{i=1}^{m}\boldsymbol{p}_i - \boldsymbol{b}(t)\right) - \dot{\boldsymbol{b}}(t)\right] - \sigma\boldsymbol{\mu}
\end{align} \tag{4.17}$$

基于方程(4.16)和(4.17), 集中式优化算法(4.15)可以写成

$$\dot{\boldsymbol{p}}_i = -\dot{\boldsymbol{\mu}} - \sigma\boldsymbol{\nabla}_{\boldsymbol{p}_i}f_i(\boldsymbol{p}_i,t) - \sigma\boldsymbol{\mu} - \boldsymbol{\nabla}_{\boldsymbol{p}_it}f_i(\boldsymbol{p}_i,t) \tag{4.18a}$$

$$\dot{\boldsymbol{\mu}} = \frac{1}{m}\left[-\sum_{i=1}^{m}\left(\sigma\boldsymbol{\nabla}_{\boldsymbol{p}_i}f_i(\boldsymbol{p}_i,t) + \boldsymbol{\nabla}_{\boldsymbol{p}_it}f_i(\boldsymbol{p}_i,t)\right) + \sigma\left(\sum_{i=1}^{m}\boldsymbol{p}_i - \boldsymbol{b}(t)\right) - \dot{\boldsymbol{b}}(t)\right] - \sigma\boldsymbol{\mu} \tag{4.18b}$$

4.3 分布式优化算法

接下来，基于集中式优化算法(4.18)，设计如下的分布式优化算法：

$$\dot{\hat{p}}_i = -\dot{\hat{\mu}}_i - \sigma\nabla_{p_i}f_i(\hat{p}_i,t) - \sigma\hat{\mu}_i - \nabla_{p_i t}f_i(\hat{p}_i,t) \tag{4.19a}$$

$$\dot{\hat{\mu}}_i = -\sigma\nabla_{p_i}f_i(\hat{p}_i,t) - \nabla_{p_i t}f_i(\hat{p}_i,t) + \sigma(\hat{p}_i - b_i(t)) - \dot{b}_i(t) - \sigma\hat{\mu}_i \\ - \alpha\sum_{j\in N_i}\text{sign}(\hat{\mu}_i - \hat{\mu}_j) \tag{4.19b}$$

其中 α 是正常数，$\hat{p}_i \in \mathbf{R}^n$ 和 $\hat{\mu}_i \in \mathbf{R}^n$ 是第 i 个智能体分别对集中式算法中 p_i 和 μ 的估计量，且 $b_i(t) \in \mathbf{R}^n$ 满足 $\sum_{i=1}^m b_i(t) = b(t)$。

引理 4.2 定义 $w_i = p_i + \mu$，$\hat{w}_i = \hat{p}_i + \hat{\mu}_i$，那么 $\lim_{t\to\infty}(\hat{w}_i - w_i) = 0$。

证明 根据 w_i 的定义，可以得到

$$\dot{w}_i = \dot{p}_i + \dot{\mu}$$

将方程(4.18a)代入上式，得到

$$\dot{w}_i = -\sigma\nabla_{p_i}f_i(p_i,t) - \sigma\mu - \nabla_{p_i t}f_i(p_i,t)$$

由于 $\nabla_{p_i}f_i(p_i,t) = p_i + \beta_i(t)$ 和 $\nabla_{p_i t}f_i(p_i,t) = \dot{\beta}_i(t)$，其中 $\beta_i(t) = -\eta_i s_i(t) - (1-\eta_i)c(t)$，因此有

$$\dot{w}_i = -\sigma w_i - \sigma\beta_i(t) - \dot{\beta}_i(t) \tag{4.20}$$

采用同样的方法，可以得到

$$\dot{\hat{w}}_i = -\sigma\hat{w}_i - \sigma\beta_i(t) - \dot{\beta}_i(t) \tag{4.21}$$

将等式(4.20)和(4.21)两边同时相减，可以得到 $\dot{\hat{w}}_i - \dot{w}_i = -\sigma(\hat{w}_i - w_i)$，因此可以推出 $\lim_{t\to\infty}(\hat{w}_i - w_i) = 0$。证毕！

> **引理 4.3**
>
> 对于集中式算法(4.18)和分布式算法(4.19), 如下的等式成立:
> $$\frac{1}{m}\sum_{i=1}^{m}\dot{\hat{\boldsymbol{\mu}}}_i - \dot{\boldsymbol{\mu}} = -\frac{\sigma}{m}\sum_{i=1}^{m}(\hat{\boldsymbol{\mu}}_i - \boldsymbol{\mu}) \tag{4.22}$$

证明 考虑 $\nabla_{\boldsymbol{p}_i}f_i(\boldsymbol{p}_i,t) = \boldsymbol{p}_i + \boldsymbol{\beta}_i(t)$ 和 $\nabla_{\boldsymbol{p}_i t}f_i(\boldsymbol{p}_i,t) = \dot{\boldsymbol{\beta}}_i(t)$, 其中 $\boldsymbol{\beta}_i(t) = -\eta_i \boldsymbol{s}_i(t) - (1-\eta_i)\boldsymbol{c}(t)$. 根据方程(4.19b), 有

$$\begin{aligned}\frac{1}{m}\sum_{i=1}^{m}\dot{\hat{\boldsymbol{\mu}}}_i &= \frac{1}{m}\sum_{i=1}^{m}\Big[-\sigma\nabla_{\boldsymbol{p}_i}f_i(\hat{\boldsymbol{p}}_i,t) - \nabla_{\boldsymbol{p}_i t}f_i(\hat{\boldsymbol{p}}_i,t) + \sigma(\hat{\boldsymbol{p}}_i - \boldsymbol{b}_i(t)) - \dot{\boldsymbol{b}}_i(t)\Big] - \frac{\sigma}{m}\sum_{i=1}^{m}\hat{\boldsymbol{\mu}}_i \\ &= -\frac{\sigma}{m}\sum_{i=1}^{m}\hat{\boldsymbol{\mu}}_i - \frac{1}{m}\sum_{i=1}^{m}\Big[\sigma\boldsymbol{\beta}_i(t) + \dot{\boldsymbol{\beta}}_i(t) + \sigma\boldsymbol{b}_i(t) + \dot{\boldsymbol{b}}_i(t)\Big]\end{aligned} \tag{4.23}$$

同时根据方程(4.18b), 有

$$\dot{\boldsymbol{\mu}} = -\sigma\boldsymbol{\mu} - \frac{1}{m}\sum_{i=1}^{m}\Big[\sigma\boldsymbol{\beta}_i(t) + \dot{\boldsymbol{\beta}}_i(t) + \sigma\boldsymbol{b}_i(t) + \dot{\boldsymbol{b}}_i(t)\Big] \tag{4.24}$$

其中用到了等式 $\boldsymbol{b}(t) = \sum_{i=1}^{m}\boldsymbol{b}_i(t)$.

进而, 根据等式(4.23)和(4.24), 得到

$$\frac{1}{m}\sum_{i=1}^{m}\dot{\hat{\boldsymbol{\mu}}}_i - \dot{\boldsymbol{\mu}} = -\frac{\sigma}{m}\sum_{i=1}^{m}(\hat{\boldsymbol{\mu}}_i - \boldsymbol{\mu})$$

证毕!

> **引理 4.4**
>
> 分布式算法(4.19)的向量 $\hat{\boldsymbol{\mu}}_i$ $(i=1,2,\cdots,m)$ 当 $t\to\infty$ 时达到一致, 即 $\lim_{t\to\infty}(\hat{\boldsymbol{\mu}}_i - \hat{\boldsymbol{\mu}}_j) = 0$, 如果参数 $\sigma > 0$ 且 α 满足
>
> $$\alpha \geqslant \kappa\sqrt{\frac{2nm}{\lambda_2(\boldsymbol{L})}} \tag{4.25}$$
>
> 其中 κ 满足
>
> $$\kappa \geqslant \sup_{t\geqslant 0}\|\sigma\boldsymbol{\beta}_i(t) + \dot{\boldsymbol{\beta}}_i(t) + \sigma\boldsymbol{b}_i(t) + \dot{\boldsymbol{b}}_i(t)\|_\infty$$

其中 $\boldsymbol{\beta}_i(t) = -\eta_i \boldsymbol{s}_i(t) - (1-\eta_i)\boldsymbol{c}(t)$.

证明 考虑 $\nabla_{\boldsymbol{p}_i} f_i(\boldsymbol{p}_i,t) = \boldsymbol{p}_i + \boldsymbol{\beta}_i(t)$ 和 $\nabla_{\boldsymbol{p}_i t} f_i(\boldsymbol{p}_i,t) = \dot{\boldsymbol{\beta}}_i(t)$，其中 $\boldsymbol{\beta}_i(t) = -\eta_i \boldsymbol{s}_i(t) - (1-\eta_i)\boldsymbol{c}(t)$. 根据方程(4.19b)，$\dot{\hat{\boldsymbol{\mu}}}_i$ 可以写成

$$\dot{\hat{\boldsymbol{\mu}}}_i = -\sigma \hat{\boldsymbol{\mu}}_i - \left(\sigma \boldsymbol{\beta}_i(t) + \dot{\boldsymbol{\beta}}_i(t) + \sigma \boldsymbol{b}_i(t) + \dot{\boldsymbol{b}}_i(t)\right) - \alpha \sum_{j \in N_i} \mathrm{sign}(\hat{\boldsymbol{\mu}}_i - \hat{\boldsymbol{\mu}}_j)$$

进而可以简写为

$$\dot{\hat{\boldsymbol{\mu}}}_i = -\sigma \hat{\boldsymbol{\mu}}_i + \hat{\boldsymbol{q}}_i(t) - \alpha \sum_{j \in N_i} \mathrm{sign}(\hat{\boldsymbol{\mu}}_i - \hat{\boldsymbol{\mu}}_j)$$

其中 $\hat{\boldsymbol{q}}_i(t) = -\sigma \boldsymbol{\beta}_i(t) - \dot{\boldsymbol{\beta}}_i(t) - \sigma \boldsymbol{b}_i(t) - \dot{\boldsymbol{b}}_i(t)$.

根据假设条件，$\hat{\boldsymbol{q}}_i(t)$ 是有界的，记 $\|\hat{\boldsymbol{q}}_i(t)\|_\infty \leqslant \kappa$. 如此引理4.1的条件满足，因此对 $i = 1, 2, \cdots, m$，当 $t \to \infty$ 时 $\hat{\boldsymbol{\mu}}_i$ 能够达到一致，即 $\lim\limits_{t \to \infty}(\hat{\boldsymbol{\mu}}_i - \hat{\boldsymbol{\mu}}_j) = 0$. 证毕！

> **定理 4.1**
> 设多智能体系统(4.19)对应的通信网络拓扑图是无向连通的，如果参数 α 满足条件(4.25)，那么智能体 i $(i=1,2,\cdots,m)$ 的状态向量 $\hat{\boldsymbol{p}}_i$ 和 $\hat{\boldsymbol{\mu}}_i$ 当 $t \to \infty$ 时收敛到时变优化问题(4.4)的最优状态和对应的拉格朗日乘子.

证明 根据引理4.4，如果参数 α 满足条件(4.25)，分布式动态系统(4.19)的状态向量 $\hat{\boldsymbol{\mu}}_i$ 当 $t \to \infty$ 时达到一致. 也就是说对 $i,j = 1,2,\cdots,m$，$\lim\limits_{t \to \infty}(\hat{\boldsymbol{\mu}}_i(t) - \hat{\boldsymbol{\mu}}_j(t)) = 0$ 成立.

根据引理4.3，等式(4.22)成立，因此 $\sum\limits_{i=1}^m (\hat{\boldsymbol{\mu}}_i - \boldsymbol{\mu})$ 指数收敛到零. 结合 $\lim\limits_{t \to \infty}(\hat{\boldsymbol{\mu}}_i - \hat{\boldsymbol{\mu}}_j) = 0$，可以得出 $\lim\limits_{t \to \infty}(\hat{\boldsymbol{\mu}}_i - \boldsymbol{\mu}) = 0$，也就是说 $\hat{\boldsymbol{\mu}}_i(t)$ $(i=1,2,\cdots,m)$ 当 $t \to \infty$ 时一致收敛到 $\boldsymbol{\mu}(t)$.

在引理4.2中，根据 $\boldsymbol{w}_i$ 和 $\hat{\boldsymbol{w}}_i$ 的定义，有 $\boldsymbol{p}_i = \boldsymbol{w}_i - \boldsymbol{\mu}$ 和 $\hat{\boldsymbol{p}}_i = \hat{\boldsymbol{w}}_i - \hat{\boldsymbol{\mu}}_i$. 由于 $\lim\limits_{t \to \infty}(\hat{\boldsymbol{w}}_i - \boldsymbol{w}_i) = 0$ 以及 $\lim\limits_{t \to \infty}(\hat{\boldsymbol{\mu}}_i - \boldsymbol{\mu}) = 0$，因此有 $\lim\limits_{t \to \infty}(\hat{\boldsymbol{p}}_i - \boldsymbol{p}_i) = 0$.

最后，因为集中式算法(4.18)的状态向量能够收敛到最优轨迹 $\boldsymbol{p}^*(t)$ 和拉格朗日乘子 $\boldsymbol{\mu}^*(t)$，所以分布式算法(4.19)的状态向量 $\hat{\boldsymbol{p}}_i(t)$ 和 $\hat{\boldsymbol{\mu}}_i(t)$ 收敛到时变优化问题(4.4)的最优状态 $\boldsymbol{p}^*(t)$ 和对应的拉格朗日乘子 $\boldsymbol{\mu}^*(t)$. 证毕！

练 习

1. 对于分布式优化算法(4.19)和图4.1的环形通信拓扑结构写出每个机器人对应的迭代算法.

2. 参考附录 A 中关于微分系统数值方法的内容，分别写出系统(4.19)对应的欧拉迭代公式和龙格-库塔迭代公式.

第 2 部分

仿 真 篇

第 5 章 ROS 2 简介

ROS 是机器人操作系统(robot operating system)的简称, 在 ROS 维基中将 ROS 定义如下①:

> **定义 5.1**
> ROS 是一个适用于机器人的开源的元操作系统. 它提供了操作系统应有的服务, 包括硬件抽象, 底层设备控制, 常用函数的实现, 进程间消息传递, 以及包管理. 它也提供用于获取、编译、编写和跨计算机运行代码所需的工具和库函数.

2007 年前后, ROS 起源于美国斯坦福大学人工智能实验室与 Willow Garage 公司的项目合作, 并于 2013 年移交给了开源机器人基金会(open source robotics foundation). 十多年来, 随着机器人与人工智能技术的发展, ROS 在学术界和工业界都得到了迅速发展.

5.1 版本选择

目前, ROS 有两个版本: ROS 1 和 ROS 2. 通常所说的 ROS 是指 ROS 1. ROS 1 与 ROS 2 的一个最大的区别是, 在 ROS 1 中有一个 master 节点, 当此节点发生故障时, 将使系统发生通信中断等严重后果, 因此不适合多机器人协作的应用场景. 此外, ROS 1 在执行命令的实时性方面也有欠缺, 难以应对实时性要求很高的问题, 如基于时变分布式优化的多机器人围堵问题. 继而, ROS 2 对 ROS 1 进行了较大的改进, 包括摒弃了 ROS 1 的 master 节点, 使得 ROS 2 的架构更加具有分布式的特点. 因此, 本书中采用 ROS 2 对算法部分进行代码实现.

ROS 2 自 2017 年发行首个版本以来, 先后发行了 8 个版本: Ardent Apalone, Bouncy Bolson, Crystal Clemmys, Dashing Diademata, Eloquent Elusor, Foxy Fitzroy, Galactic Geochelone, Humble Hawksbill, 本书中选用最新的版本为 Humble Hawksbill.

ROS 2 支持多种操作系统, 本书使用的 ROS 2 应用程序开发环境如下:

- 操作系统: Ubuntu 22.04 Jammy Jellyfish
- ROS 2: Humble Hawksbill

---

① https://wiki.ros.org.

Ubuntu 操作系统可以采用实体机运行、虚拟机运行，或者 WSL(windows subsystem for linux)。本书不介绍操作系统的安装，采用的是 WSL 环境下的 Ubuntu 操作系统。以下介绍在 Ubuntu 系统中如何安装 ROS 2。

5.2 安装并测试 ROS 2

在 Ubuntu 22.04 系统中安装 ROS 2 Humble，分为如下几个步骤[①]：

1. 设置编码[②]

```
$ sudo apt update && sudo apt install locales
$ sudo locale-gen en_US en_US.UTF-8
$ sudo update-locale LC_ALL=en_US.UTF-8 LANG=en_US.UTF-8
$ export LANG=en_US.UTF-8
```

2. 添加库

```
$ sudo apt update && sudo apt install curl gnupg lsb-release
$ sudo curl -sSL https://raw.githubusercontent.com/ros/rosdistro/master/ros.
  key -o /usr/share/keyrings/ros-archive-keyring.gpg
$ echo "deb [arch=$(dpkg --print-architecture) signed-by=/usr/share/keyrings
  /ros-archive-keyring.gpg] http://packages.ros.org/ros2/ubuntu $(source /
  etc/os-release && echo $UBUNTU_CODENAME) main" | sudo tee /etc/apt/
  sources.list.d/ros2.list > /dev/null
```

3. 安装 ROS 2 Humble

```
$ sudo apt update
$ sudo apt upgrade
$ sudo apt install ros-humble-desktop
```

4. 环境设置

```
$ echo "source /opt/ros/humble/setup.bash" >> ~/.bashrc
$ source ~/.bashrc
```

[①] ROS 2 Humble 的安装步骤可以参考https://docs.ros.org/en/humble/Installation/Ubuntu-Install-Debians.html。

[②] 本书中以 $ 符号开始的命令表示是在 Ubuntu 终端窗口中执行的命令。

5. 测试

打开两个终端窗口,分别运行如下两条命令:

```
$ ros2 run turtlesim turtlesim_node
$ ros2 run turtlesim turtle_teleop_key
```

如果 ROS 2 安装成功的话,可以通过键盘的方向键控制小海龟进行移动 (如图5.1所示).

图 5.1　键盘控制小海龟移动

5.3　创建并初始化本书的工作目录

ROS 2 使用 colcon build 进行系统构建. 为了使用它, 需要创建并初始化工作目录. 如下所示, 创建本书的工作目录:

```
$ mkdir -p ~/book_ws/src
```

运行上述命令之后, 会在用户目录下生成一个 book_ws 的工作目录, 以及在目录 book_ws 下生成 src 目录. 本书中的程序源文件都将保存在此 src 目录下. 接下来, 使用 colcon build 命令进行程序构建.

```
$ cd ~/book_ws
$ colcon build
```

如果构建没有问题, 运行 ls 命令, 可以看到在 book_ws 目录下生成了新的 build, install 和 log 目录. colcon 的构建系统的相关文件保存在 build 目录中, 构建后的安装文件保存在 install 目录中, 日志文件保存在 log 目录中.

运行如下命令将与 colcon 构建系统相关的环境文件写入设置文件中并使之生效:

```
$ echo "source ~/book_ws/install/setup.bash" >> ~/.bashrc
$ source ~/.bashrc
```

5.4 创建第一个功能包

执行如下命令创建一个名称为 chapter5_4 的功能包:

```
$ cd ~/book_ws/src
$ ros2 pkg create --build-type ament_python chapter5_4
```

接下来,构建这个包:

```
$ cd ~/book_ws
$ colcon build
```

也可以用如下命令只对一个功能包进行构建:

```
$ cd ~/book_ws
$ colcon build --packages-select chapter5_4
```

生成安装文件的源文件:

```
$ . install/setup.bash
```

以上构建的功能包没有实现任何功能,后续将通过介绍 ROS 2 的通信机制构建新的功能包. 接下来,介绍 ROS 2 的通信接口(interface). 通信是 ROS 2 的一个核心功能和概念,主要包括三种类型的接口: 消息、服务和动作.

5.5 消 息

消息 (message) 通信是指发送信息的发布者和接收信息的订阅者以话题(topic)的形式发送和接收消息. 接收消息的订阅者节点接收的是与发布者节点发布的具有相同话题名称的消息. 基于这个消息, 订阅者节点直接连接到发布者节点并接收消息.

如图5.2所示, 机器人 A 在移动过程中将自身的坐标位置通过话题 "odom" 的形式发布给机器人 B 和机器人 C, 以此告诉对方自己的位置信息. 消息通信的传播是单向的, 适用于需要连续发送消息的传感器数据. 另外, 单个发布者可以与多个订阅者进行通信.

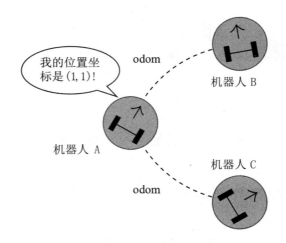

图 5.2　机器人 A 将里程计信息通过话题 "odom" 同时发布给机器人 B 和机器人 C

在 book_ws/src 目录下构建一个新的功能包:

```
$ cd ~/book_ws/src
$ ros2 pkg create --build-type ament_cmake chapter5_interfaces
```

注意到这里构建的 chapter5_interfaces 是一个 CMake 的功能包, 目前还没有办法在纯 Python 的功能包中生成 .msg 或 .srv 文件. 但是我们可以在 CMake 包中创建一个自定义接口, 然后在 Python 节点中使用它.

接下来, 建立一个名称为 msg 的目录, 用来存放将要自定义的消息文件:

```
$ cd ~/book_ws/src/chapter5_interfaces
$ mkdir msg
```

在 chapter5_interfaces/msg 目录下新建一个名称为 Num.msg 的消息文件, 其内容如下:

```
1  int64 a
2  int64 b
```

这个文件可以用来传输两个数据类型是 64 位整数的变量 a 和 b.

为了在特定的编程语言中能够使用上述自定义的消息类型, 在 chapter5_interfaces 目录下的 CMakeLists.txt 文件的 ament_package() 内容前添加如下代码:

```
1  find_package(rosidl_default_generators REQUIRED)
2  rosidl_generate_interfaces(${PROJECT_NAME}
3    "msg/Num.msg"
4  )
```

接下来, 添加依赖项. 在 chapter5_interfaces 目录下的 package.xml 文件的 </package> 内容前添加如下代码:

```
1  <build_depend>rosidl_default_generators</build_depend>
2  <exec_depend>rosidl_default_runtime</exec_depend>
3  <member_of_group>rosidl_interface_packages</member_of_group>
```

构建 chapter5_interfaces 功能包，运行如下命令：

```
$ cd ~/book_ws
$ colcon build --packages-select chapter5_interfaces
```

接下来，参考第5.4节的方法创建一个名称为 chapter5_5 的功能包，但是这里将实现两个整数的加法运算。

执行如下命令创建一个名称为 chapter5_5 的功能包：

```
$ cd ~/book_ws/src
$ ros2 pkg create --build-type ament_python chapter5_5
```

然后构建这个包：

```
$ cd ~/book_ws
$ colcon build --packages-select chapter5_5
$ . install/setup.bash
```

1. 创建发布者节点

通过如下命令创建发布者节点的 Python 文件[①]：

```
$ cd ~/book_ws/src/chapter5_5/chapter5_5
$ sudo nano topic_publisher_Num.py
```

在 topic_publisher_Num.py 文件中添加如下代码：

```
1   #!/usr/bin/env python3
2   import rclpy
3   from rclpy.node import Node
4   import random
5   from chapter5_interfaces.msg import Num
6
7   class Publisher(Node):
8       def __init__(self):
9           super().__init__('publisher')
10          self.publisher = self.create_publisher(Num, 'two_integers', 10)
11          timer_period = 0.5
12          self.random_numbers = self.create_timer(timer_period, self.
                random_numbers_callback)
```

---

① nano 是 Ubuntu 系统的一个小巧而方便的文本编辑器。

```
13
14      def random_numbers_callback(self):
15          msg = Num()
16          msg.a = random.randint(0,100)
17          msg.b = random.randint(0,100)
18          self.publisher.publish(msg)
19          print("随机生成两个整数: %d, %d" % (msg.a, msg.b))
20
21  def main():
22      rclpy.init()
23      publisher = Publisher()
24      rclpy.spin(publisher)
25      publisher.destroy_node()
26      rclpy.shutdown()
27
28  if __name__ == '__main__':
29      main()
```

这是第一个程序,接下来详细介绍它的含义:
- 开始的第 1 行说明这个程序会被看作 Python 3 的程序脚本;
- 第 2~5 行加载了运行此程序需要用到的包,其中 rclpy 是 ROS 2 的核心 Python 库, rclpy.node 的 Node 将用于自定义的类,以继承 Node 类的所有功能, random 用于程序中生成随机数,而 Num 正是前面自定义的消息类型;
- 第 7~19 行定义了一个节点类 Publisher() 用于创建发布者,主要包括以下内容:
    - 第 9 行初始化发布者节点名称为 publisher;
    - 第 10 行使用 Node 类的 create_publisher 方法创建一个发布者,使用自定义的消息类型 Num, 发布的话题名称是 two_integers, 发布的消息队列大小是 10;
    - 第 11 行定义一个常量表示消息发布的周期为 0.5 秒,也就是发布频率是 2;
    - 第 12 行通过 Node 类的 create_timer 方法启动一个计时器,并按照周期 0.5 秒调用回调函数 random_numbers_callback();
    - 第 14~19 行定义了回调函数 random_numbers_callback(), 用自定义的消息类型 Num 作为通信接口,具体化为变量 msg, 并将其包含的两个消息变量 a 和 b 都取 0 到 100 之间的随机整数,然后用 publish 方法将这两个随机整数发布出去,并在屏幕上打印这两个整数.
- 第 21~26 行是主程序,其中 rclpy.init() 是使用 ROS 2 时需要对节点进行初始化,然后,将前面定义的类 Publisher() 实例化生成 publisher 节点,并使用 spin 方法让这个节点一直处于运行状态直至同时按下键盘上的 'ctrl' 和 'c' 键,然后删除这个节点并关闭所有执行器.

接下来，为了使前面定义的脚步文件可执行，还需要打开 chapter5_5 目录下的 setup.py 文件，将 entry_points 域修改为如下形式：

```
entry_points={
    'console_scripts': [
            'talker = chapter5_5.topic_publisher_Num:main',
    ],
},
```

即添加了第 3 行代码，其中的 talker 是可以自定义的名称，也是以后在调用或者执行这个节点脚本时可以使用的名称.

2. 创建订阅者节点

完成了前面的发布者节点的脚本程序，还需要编写订阅者节点的脚本程序. 接下来，通过如下命令创建订阅者节点的 Python 文件，用于实现求解两个整数的和：

```
$ cd ~/book_ws/src/chapter5_5/chapter5_5
$ sudo nano topic_subscriber_Add.py
```

在 topic_subscriber_Add.py 文件中添加如下代码：

```python
#!/usr/bin/env python3
import rclpy
from rclpy.node import Node
from chapter5_interfaces.msg import Num

class SubscriberAdd(Node):
    def __init__(self):
        super().__init__('subscriber_add')
        self.create_subscription(Num, 'two_integers', self.add_callback, 10)

    def add_callback(self, msg):
        add_a_b = msg.a + msg.b
        print("求两个整数的和: %d + %d = %d" % (msg.a, msg.b, add_a_b))

def main():
    rclpy.init()
    adder = SubscriberAdd()
    rclpy.spin(adder)
    adder.destroy_node()
    rclpy.shutdown()

if __name__ == '__main__':
    main()
```

仅对这个程序中与前一个程序区别较大的代码进行说明, 相似的代码不再赘述:
- 第 6~13 行定义了一个节点类 SubscriberAdd() 用于创建订阅者, 主要包括以下内容:
  - 第 8 行初始化订阅者节点名称为 subscriber_add;
  - 第 9 行使用 Node 类的 create_subscription 方法订阅话题 two_integers, 注意这里的消息类型也是 Num, 需要与发布者一致, 话题名称也需要与发布者的话题名称一致, 同时订阅者调用回调函数 add_callback() 对订阅的数据进行处理, 订阅的消息队列大小设置为 10;
  - 第 11~13 行定义了回调函数 add_callback(), 实现对订阅的两个整数进行求和运算, 并在屏幕上打印运算结果.
- 第 15~20 行是主程序, 将前面定义的类 SubscriberAdd() 实例化生成 adder 节点.

同样地, 为了使定义的脚步文件可执行, 需要打开 chapter5_5 目录下的 setup.py 文件, 将 entry_points 域修改为如下形式:

```
1  entry_points={
2      'console_scripts': [
3              'talker = chapter5_5.topic_publisher_Num:main',
4              'adder = chapter5_5.topic_subscriber_Add:main',
5      ],
6  },
```

即添加一个名称为 adder 的可执行节点.

接下来, 重新构建 chapter5_5 功能包, 运行如下命令:

```
$ cd ~/book_ws
$ colcon build --packages-select chapter5_5
```

打开两个终端, 分别输入如下两条命令进行测试:

```
$ ros2 run chapter5_5 talker
$ ros2 run chapter5_5 adder
```

注 这里的 ros2 run 是 ROS 2 运行单个节点的命令, 其格式是 "ros2 run 功能包 可执行节点". 在这个例子中, 功能包是 chapter5_5, 可执行节点是 talker 和 adder, 即是前面在 setup.py 文件中自定义的两个节点.

运行结果如图5.3所示, 左侧窗口显示的是发布者节点 "publisher" 随机产生的两个整数, 右侧窗口是订阅者节点 "subscriber_add" 对这两个随机整数进行求和运算.

3. 节点和话题分析

接下来, 对上述 talker 和 adder 的节点信息进行分析. 同样地, 打开两个终端, 分别运行如下两个节点:

```
lqs@ubuntu:~$ ros2 run chapter5_5 talker          lqs@ubuntu:~$ ros2 run chapter5_5 adder
随机生成两个整数: 62, 50                          求两个整数的和: 62 + 50 = 112
随机生成两个整数: 89, 29                          求两个整数的和: 89 + 29 = 118
随机生成两个整数: 0, 60                           求两个整数的和: 0 + 60 = 60
随机生成两个整数: 34, 32                          求两个整数的和: 34 + 32 = 66
随机生成两个整数: 63, 90                          求两个整数的和: 63 + 90 = 153
随机生成两个整数: 62, 32                          求两个整数的和: 62 + 32 = 94
随机生成两个整数: 54, 60                          求两个整数的和: 54 + 60 = 114
随机生成两个整数: 42, 44                          求两个整数的和: 42 + 44 = 86
随机生成两个整数: 32, 33                          求两个整数的和: 32 + 33 = 65
随机生成两个整数: 15, 8                           求两个整数的和: 15 + 8 = 23
随机生成两个整数: 95, 23                          求两个整数的和: 95 + 23 = 118
随机生成两个整数: 12, 70                          求两个整数的和: 12 + 70 = 82
随机生成两个整数: 25, 83                          求两个整数的和: 25 + 83 = 108
```

图 5.3　对两个随机整数进行求和运算

```
$ ros2 run chapter5_5 talker
$ ros2 run chapter5_5 adder
```

打开第三个终端, 运行如下命令查看节点信息:

```
$ ros2 node list
```

终端将显示/publisher 和/subscriber_add 这两个节点, 其中前者是在 topic_publisher_Num.py 文件中定义的节点名称, 而后者是在 topic_subscriber_Add.py 文件中定义的节点名称.

在第三个终端中运行如下命令查看话题信息:

```
$ ros2 topic list
```

终端将显示/parameter_events、/rosout 和/two_integers 这三个话题, 其中前两个话题是 ROS 2 系统自带的参数和输出话题, 第三个正是在 topic_publisher_Num.py 文件中发布的话题.

接下来, 运行如下命令观察/two_integers 话题与节点的发布和订阅关系:

```
$ rqt_graph
```

从图5.4可以看出, /publisher 节点将/two_integers 话题传输给了/subscriber_add 节点.

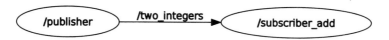

图 5.4　发布者、订阅者和话题之间的关系图

5.6 服 务

服务 (service) 通信是指请求服务的服务客户端与负责服务响应的服务服务器之间的同步双向服务消息通信. 如图5.5所示, 机器人 A 向服务器询问自己的移动目标, 服务器通过整体规划, 将机器人 A 的移动目标发送给对方.

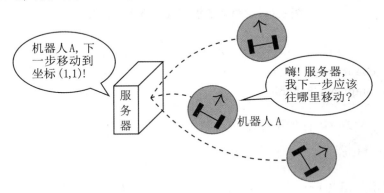

图 5.5 机器人与服务器之间通过服务话题进行双向通信

将服务分成服务服务器和服务客户端, 其中服务服务器只在有请求 (request) 的时候才响应(response), 而服务客户端会在发送请求后接收响应. 与消息通信不同, 服务通信是一次性的. 因此, 当服务的请求和响应完成时, 两个连接的节点将被断开. 该服务通常被用作请求机器人执行特定操作时使用的命令, 或者用于根据特定条件需要产生事件的节点. 由于它是一次性的通信方式, 又因为它在网络上的负载很小, 所以它也被用作代替消息的通信方式, 因此是一种非常有用的通信手段.

在上一节创建的功能包 chapter5_interfaces 下建立一个名称为 srv 的目录, 用来存放将要自定义的服务文件:

```
$ cd ~/book_ws/src/chapter5_interfaces
$ mkdir srv
```

在 chapter5_interfaces/srv 目录下新建一个名称为 Calculator.srv 的服务文件, 内容如下:

```
1  int64 a
2  int64 b
3  ---
4  int64[] result
```

这个文件可以用来传输两个数据类型是 64 位整数的自定义变量 a 和 b, 以及一个 64 位整数数组的服务响应 result. 注意这里的 int64[ ] 表示整数数组类型, 即可以同时传输多个整数.

和前面一样，为了在特定的编程语言中能够使用上述自定义的服务文件，修改chapter5_interfaces 目录下的 CMakeLists.txt 文件并添加 Calculator.srv 服务类型，添加后的代码如下所示：

```
1  find_package(rosidl_default_generators REQUIRED)
2  rosidl_generate_interfaces(${PROJECT_NAME}
3    "msg/Num.msg"
4    "srv/Calculator.srv"
5  )
```

重新构建 chapter5_interfaces 功能包，运行如下命令：

```
$ cd ~/book_ws
$ colcon build --packages-select chapter5_interfaces
```

接下来，和上一节的方法类似，创建一个名称为 chapter5_6 的功能包，用自定义的服务接口同时实现两个整数的加法和减法运算。

执行如下命令创建一个名称为 chapter5_6 的功能包：

```
$ cd ~/book_ws/src
$ ros2 pkg create --build-type ament_python chapter5_6
```

接下来，构建这个包：

```
$ cd ~/book_ws
$ colcon build --packages-select chapter5_6
$ . install/setup.bash
```

1. 创建服务器节点

通过如下命令创建服务器节点的 Python 文件：

```
$ cd ~/book_ws/src/chapter5_6/chapter5_6
$ sudo nano calculator_server.py
```

在 calculator_server.py 文件中添加如下代码：

```
1  #!/usr/bin/env python3
2  import rclpy
3  from rclpy.node import Node
4  from chapter5_interfaces.srv import Calculator
5
6  class CalculatorService(Node):
7      def __init__(self):
8          super().__init__('calculator_server')
9          self.srv = self.create_service(Calculator, 'calculator_ints', self.
              add_sub_callback)
```

```python
10
11      def add_sub_callback(self, request, response):
12          add_a_b = request.a + request.b
13          sub_a_b = request.a - request.b
14          response.result = [add_a_b, sub_a_b]
15          print("服务器运行计算: %d + %d = %d, %d - %d = %d" % (request.a,
                  request.b, add_a_b, request.a, request.b, sub_a_b))
16          return response
17
18  def main():
19      rclpy.init()
20      calculator_service = CalculatorService()
21      rclpy.spin(calculator_service)
22      calculator_service.destroy_node()
23      rclpy.shutdown()
24
25  if __name__ == '__main__':
26      main()
```

下面解释上述代码的主要内容:

- 第 2~4 行导入需要用到的包,其中包括前面自定义的服务类型 Calculator;
- 第 6~16 行定义了一个节点类 CalculatorService() 用于创建服务器节点,主要包括以下内容:
  - 第 8 行初始化服务器节点名称为 calculator_server;
  - 第 9 行使用 Node 类的 create_service 方法创建一个服务器,使用自定义的服务类型 Calculator,发布的服务名称是 calculator_ints,并调用回调函数 add_sub_callback() 进行整数运算;
  - 第 11~16 行定义了回调函数 add_sub_callback(),当服务器收到客户端的请求时,计算两个整数的和与差,其中变量 request 用于存储客户端发来的请求,变量 response 用于存储服务器的响应,同时服务器将计算过程打印在屏幕上,并将 response 作为回调函数的返回值.
- 第 18~23 行是主程序,其内容和功能与前述代码类似,这里不再对其进行解释.

2. 创建客服端节点

通过如下命令创建客服端节点的 Python 文件:

```
$ cd ~/book_ws/src/chapter5_6/chapter5_6
$ sudo nano calculator_client.py
```

在 calculator_client.py 文件中添加如下代码:

```python
#!/usr/bin/env python3
import rclpy
from rclpy.node import Node
import random
from chapter5_interfaces.srv import Calculator

class CalculatorClient(Node):
    def __init__(self):
        super().__init__('calculator_client')
        self.cli = self.create_client(Calculator, 'calculator_ints')
        while not self.cli.wait_for_service(timeout_sec=1.0):
            self.get_logger().info('找不到服务器，请等待 ...')
        self.req = Calculator.Request()

    def send_request(self):
        self.req.a = random.randint(0,100)
        self.req.b = random.randint(0,100)
        self.future = self.cli.call_async(self.req)

def main():
    rclpy.init()
    calculator_client = CalculatorClient()
    calculator_client.send_request()
    rclpy.spin_once(calculator_client)
    if calculator_client.future.done():
        response = calculator_client.future.result()
        print("整数%d和%d的和与差是%d与%d" % (calculator_client.req.a,
            calculator_client.req.b, response.result[0], response.result[1]))
    calculator_client.destroy_node()
    rclpy.shutdown()

if __name__ == '__main__':
    main()
```

下面解释代码的主要部分：

- 第 2~5 行导入需要用到的包，其中包括用于生成随机数的 random 和前面自定义的服务类型 Calculator；
- 第 7~18 行定义了一个节点类 CalculatorClient() 用于创建客户端节点，主要包括以下内容：
    - 第 9 行初始化客户端节点名称为 calculator_client；

- 第 10 行使用 Node 类的 create_client 方法创建一个客户端，使用自定义的服务类型 Calculator，使用的服务名称是 calculator_ints，和服务器的服务名称保持一致;
- 第 11~12 行给出了程序报错信息，即如果客户端无法和服务器通信，则在屏幕上显示报错信息，并将该信息写入日志文件中;
- 第 13 行初始化客户端的请求变量;
- 第 15~18 行定义了函数 send_request() 实现随机生成两个 0 到 100 之间的整数，并将其作为请求发送给服务器.
- 第 20~29 行是主程序，主要包括以下内容:
  - 第 22 行将前面定义的类 CalculatorClient() 实例化生成 calculator_client 节点;
  - 第 23 行调用函数 send_request() 发送请求给服务器，即发送两个随机生成的整数给服务器，同时请求服务器对这两个整数进行处理，即服务器会计算这两个整数的和与差;
  - 第 24 行使用 spin_once 方法使客户端节点仅运行一次;
  - 第 25~27 行，如果客户端的请求和服务器端的响应能够成功(节点通过 future.done 方法检查是否收到服务器的响应)，则通过 future.result 方法返回服务器的结果并存储在变量 response 中，即 response 中包含了前面定义的服务类型接口 Calculator 的 result 的内容，同时将相应的计算结果打印在屏幕上.

注 这里的 spin_once 与之前使用的 spin 不同，前者表示节点只执行一项工作或等待超时结束，而后者表示节点会一直执行类中所定义的工作.

然后在 chapter5_6 目录下的 setup.py 文件中修改 entry_points 域为如下形式:

```
1  entry_points={
2      'console_scripts': [
3              'server = chapter5_6.calculator_server:main',
4              'client = chapter5_6.calculator_client:main',
5      ],
6  },
```

构建 chapter5_6 功能包，运行如下命令:

```
$ cd ~/book_ws
$ colcon build --packages-select chapter5_6
```

打开两个终端，分别输入如下两条命令进行测试:

```
$ ros2 run chapter5_6 server
$ ros2 run chapter5_6 client
```

运行结果如图5.6所示，左侧窗口显示的是服务器计算的结果，右侧窗口是客服端收到的运算结果．

图 5.6 对两个随机整数进行求和与作差运算

5.7 动 作

动作 (action) 通信主要针对如下的情形进行处理：服务器收到客户端的请求后做出响应所需的时间较长，且需要中途反馈值．动作通信与服务通信非常相似，但动作通信比服务通信多了反馈 (feedback)．如图5.7所示，当机器人 A 询问机器人 B 是否到达目标位置时，机器人 B 在移动过程中会多次告诉机器人 A 自己的位置，并在到达目标位置时通知机器人 A．与服务通信不同，动作通信常用于指导复杂的机器人任务，例如发送一个目标值之后，还可以在任意时刻发送取消目标的命令．

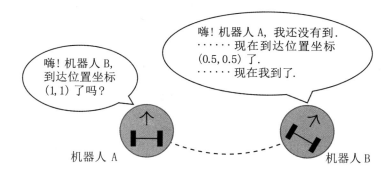

图 5.7 机器人 A 与机器人 B 之间通过动作通信进行队形协调

在前面创建的功能包 chapter5_interfaces 下建立一个名称为 action 的目录，用来存放将要自定义的动作通信文件：

```
$ cd ~/book_ws/src/chapter5_interfaces
$ mkdir action
```

在 chapter5_interfaces/action 目录下新建一个名称为 CountUntil.action 的动作文件，文件内容如下：

```
1  int64 max_number
2  ---
3  int64 total
4  ---
```

```
5    int64 count
```

这里的 max_number 用来记录完成求和的最大整数，total 输出求和结果，count 记录当前所累加的整数。

和前面一样，为了在特定的编程语言中能够使用上述自定义的动作类型，修改 chapter5_interfaces 目录下的 CMakeLists.txt 文件并添加 CountUntil.action 接口文件，添加后的代码如下所示：

```
1    find_package(rosidl_default_generators REQUIRED)
2    rosidl_generate_interfaces(${PROJECT_NAME}
3      "msg/Num.msg"
4      "srv/Calculator.srv"
5      "action/CountUntil.action"
6    )
```

重新构建 chapter5_interfaces 功能包，运行如下命令：

```
$ cd ~/book_ws
$ colcon build --packages-select chapter5_interfaces
```

接下来，和上一节的方法类似，创建一个名称为 chapter5_7 的功能包，使用自定义的动作类型实现整数序列的加法运算。

执行如下命令创建一个名称为 chapter5_7 的功能包：

```
$ cd ~/book_ws/src
$ ros2 pkg create --build-type ament_python chapter5_7
```

接下来，构建这个包：

```
$ cd ~/book_ws
$ colcon build --packages-select chapter5_7
$ . install/setup.bash
```

1. 创建动作服务器节点

通过如下命令创建动作服务器节点的 Python 文件：

```
$ cd ~/book_ws/src/chapter5_7/chapter5_7
$ sudo nano action_server.py
```

在 action_server.py 文件中添加如下代码：

```
1    #!/usr/bin/env python3
2    import rclpy
3    from rclpy.node import Node
4    from rclpy.action import ActionServer
```

```
 5    import time
 6    from chapter5_interfaces.action import CountUntil
 7
 8    class CountUntilActionServer(Node):
 9        def __init__(self):
10            super().__init__('action_server')
11            self.action_server = ActionServer(self, CountUntil, 'countuntil', self
                  .count_callback)
12
13        def count_callback(self, goal_handle):
14            feedback = CountUntil.Feedback() # feedback = goal_handle.feedback 待
                  验证
15            result = CountUntil.Result() # result = goal_handle.result 待验证
16            feedback.count = 0
17            result.total = 0
18
19            for i in range(1, goal_handle.request.max_number):
20                feedback.count = i
21                result.total = result.total + i
22                print('当前累加到: %d' % feedback.count)
23                goal_handle.publish_feedback(feedback)
24                time.sleep(0.5)
25
26            goal_handle.succeed()
27            return result
28
29    def main():
30        rclpy.init()
31        action_server = CountUntilActionServer()
32        rclpy.spin(action_server)
33        action_server.destroy_node()
34        rclpy.shutdown()
35
36    if __name__ == '__main__':
37        main()
```

下面解释上述代码的主要内容:

- 第 2~6 行导入需要用到的包,其中包括 ActionServer 包用于生成动作服务器, time 包用于后面将使用 time.sleep() 设置每一轮计算的周期,以及自定义的动作类型 CountUntil;
- 第 8~27 行定义了一个节点类 CountUntilActionServer() 用于创建动作服务器节

点，主要包括以下内容：
- 第 10 行初始化动作服务器节点名称为 action_server；
- 第 11 行使用动作库的 ActionServer 方法创建一个动作服务器，使用自定义的动作类型 CountUntil，定义的动作名称是 countuntil，并调用回调函数 count_callback()；
- 第 13~27 行定义了回调函数 count_callback()，其内容解释如下：
  - 第 14 行通过 Feedback 方法定义了变量 feedback 用于存储反馈值；
  - 第 15 行通过 Result 方法定义了变量 result 用于存储结果值；
  - 第 16~17 行分别初始化 feedback.count 和 result.total 的值为零；
  - 第 19~24 行对整数进行累加，其中第 19 行的迭代上界 request.max_number 取自动作客户端的请求值，第 22 行将反馈值打印在动作服务器的屏幕上，第 23 行将反馈发布给动作客服端，第 24 行设置每次累加间隔时间是 0.5 秒；
  - 第 26 行使用 succeed 方法表示目标成功；
  - 第 27 行将动作服务器的结果作为回调函数的返回值.
- 第 29~34 行是主程序，主要包括以下内容：
  - 第 31 行将前面定义的类 CountUntilActionServer() 实例化生成 action_server 节点. 其他代码与之前的程序类似，这里不再赘述.

2. 创建动作客服端节点

同样地，通过如下命令创建动作客服端节点的 Python 文件：

```
$ cd ~/book_ws/src/chapter5_7/chapter5_7
$ sudo nano action_client.py
```

在 action_client.py 文件中添加如下代码：

```
1   #!/usr/bin/env python3
2   import rclpy
3   from rclpy.node import Node
4   from rclpy.action import ActionClient
5   from chapter5_interfaces.action import CountUntil
6
7   class CountUntilActionClient(Node):
8       def __init__(self):
9           super().__init__('action_client')
10          self.action_client = ActionClient(self, CountUntil, 'countuntil')
11
12      def send_goal(self, max_number):
13          self.goal = CountUntil.Goal()
14          self.goal.max_number = max_number
```

```python
15          self.action_client.wait_for_server()
16          self.send_goal_future = self.action_client.send_goal_async(self.goal,
                self.feedback_callback)
17          self.send_goal_future.add_done_callback(self.goal_response_callback)
18
19      def feedback_callback(self, feedback_msg):
20          feedback = feedback_msg.feedback
21          print('当前计数: %d' % feedback.count)
22
23      def goal_response_callback(self, future):
24          goal_handle = future.result()
25          self.get_result_future = goal_handle.get_result_async()
26          self.get_result_future.add_done_callback(self.get_result_callback)
27
28      def get_result_callback(self, future):
29          result = future.result().result
30          print('1到%d的整数累加和是: %d' % (self.goal.max_number-1, result.
                total))
31          rclpy.shutdown()
32
33  def main():
34      rclpy.init()
35      action_client = CountUntilActionClient()
36      action_client.send_goal(11)
37      rclpy.spin(action_client)
38
39  if __name__ == '__main__':
40      main()
```

代码中定义了类 CountUntilActionClient() 用于实现动作客户端的操作, 其中函数 send_goal() 实现将目标请求发送给服务器, 并通过调用 feedback_callback() 回调函数在客户端的屏幕打印当前计数, 函数 goal_response_callback() 获取当前的计算结果, 并通过调用函数 get_result_callback() 在屏幕上打印最终的计算结果.

下面解释上述代码的主要内容:

- 第 2~5 行导入需要用到的包, 其中包括 ActionClient 包用于生成动作客户端, 以及自定义的动作类型 CountUntil;
- 第 7~31 行定义了一个节点类 CountUntilActionClient() 用于创建动作客户端节点, 主要包括以下内容:
  - 第 9 行初始化动作客户端节点名称为 action_client;
  - 第 10 行使用动作库的 ActionClient 方法创建一个动作客户端, 使用自定义

的动作类型 CountUntil, 并使用和动作服务器同样的动作名称 countuntil;
- 第 12~17 行定义了动作客户端的请求函数 send_goal(), 同时调用回调函数 feedback_callback() 获取动作服务器的反馈, 以及回调函数 goal_response_callback() 获取动作服务器的结果;
- 第 19~21 行定义了回调函数 feedback_callback() 获取动作服务器的反馈, 并将其打印在屏幕上;
- 第 23~26 行定义了回调函数 goal_response_callback() 获取动作服务器的中间计算结果, 并通过调用回调函数 get_result_callback() 获取动作服务器的最终计算结果;
- 第 28~31 行定义了回调函数 get_result_callback() 获取动作服务器的最终计算结果, 并将其打印在屏幕上.
- 第 33~37 行是主程序, 主要包括以下内容:
    - 第 35 行将前面定义的类 CountUntilActionClient() 实例化生成 action_client 节点;
    - 第 36 行使用 send_goal 方法将最大计数发送给动作服务器, 这里设置的是 11, 即动作客户端请求动作服务器计算 1 到 10 的累加和.

打开 chapter5_7 目录下的 setup.py 文件, 将 entry_points 域修改为如下形式:

```
entry_points={
    'console_scripts': [
            'action_server = chapter5_7.action_server:main',
            'action_client = chapter5_7.action_client:main',
    ],
},
```

构建 chapter5_7 功能包, 运行如下命令:

```
$ cd ~/book_ws
$ colcon build --packages-select chapter5_7
```

接下来, 运行上述代码进行测试. 打开两个终端, 分别输入如下两条命令进行测试:

```
$ ros2 run chapter5_7 action_server
$ ros2 run chapter5_7 action_client
```

运行结果如图5.8所示, 左侧窗口显示的是动作服务器计算的过程, 右侧窗口是动作客服端获得的反馈和最终的计算结果.

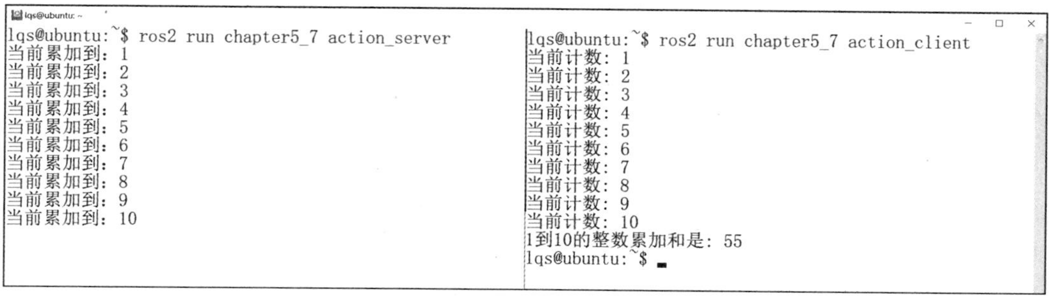

```
lqs@ubuntu:~$ ros2 run chapter5_7 action_server     lqs@ubuntu:~$ ros2 run chapter5_7 action_client
当前累加到: 1                                        当前计数: 1
当前累加到: 2                                        当前计数: 2
当前累加到: 3                                        当前计数: 3
当前累加到: 4                                        当前计数: 4
当前累加到: 5                                        当前计数: 5
当前累加到: 6                                        当前计数: 6
当前累加到: 7                                        当前计数: 7
当前累加到: 8                                        当前计数: 8
当前累加到: 9                                        当前计数: 9
当前累加到: 10                                       当前计数: 10
                                                    1到10的整数累加和是: 55
                                                    lqs@ubuntu:~$
```

图 5.8　计算 1 到 10 的所有整数的累加和

练　习

1. 参考第 5.5 节实现两个随机整数求和运算的代码，练习实现两个随机整数作差的运算.
2. 参考第 5.6 节实现两个随机整数求和与作差运算的代码，练习实现两个随机整数的四则运算.
3. 参考第 5.5 节的代码，假设发布者的发布频率是 10，订阅者的订阅频率是 5，试编写程序实现上述发布和订阅任务.

第 6 章   ROS 2 坐标变换

在第3章介绍了向量的几何变换,包括平移、旋转和伸缩变换. 机器人在空间中的坐标变换和向量的几何变换类似. 但是对于刚体来说,通常只有平移变换和旋转变换. 坐标变换无论对于单个机器人还是多个机器人,都具有重要的意义. 例如,对于人形机器人,机器人的手部相对于肘部,肘部相对于肩部,以此类推,通过坐标变换,可以得到机器人手部相对于双脚中心的坐标变换,从而可以实现机器人的手脚协调,以完成复杂的任务规划. 此外,在多机器人编队任务中,需要通过机器人之间的相对位置来确定机器人在统一坐标系下的位置,这可以通过适当的坐标变换实现坐标系的统一. 在机器人编程中,经过坐标变换的机器人的状态和位置是非常重要的,在 ROS 2 中以坐标变换 (transform) 来表达. 在多机器人跟随或者编队任务中,也需要考虑各个机器人坐标系的变换关系,这也可以通过 ROS 2 的坐标变换来实现.

6.1   实现小海龟的领导–跟随任务

在这一节中,将实现三只小海龟的领导–跟随任务. 其中小海龟 turtle1 可以通过键盘控制其移动,小海龟 turtle2 与 turtle1 保持相对位置坐标 $(-1,1)$,即 turtle2 在 turtle1 的左后方. 小海龟 turtle3 与 turtle1 保持相对位置坐标 $(-1,-1)$,即 turtle3 在 turtle1 的右后方. 如图6.1所示,turtle1 是领导者,turtle2 向自己的目标点 t_1 移动,turtle3 向自己的目标点 t_2 移动. 为了实现这个任务,可以通过 ROS 2 发布两个相对 turtle1 的静态坐标变换,然后让 turtle2 和 turtle3 追踪这两个静态坐标,以达到领导–跟随的效果.

6.1.1   小海龟状态发布

首先,介绍如何将机器人(这里先用小海龟)的状态在世界坐标系下发布出去,以用于实现后续的跟踪任务.

执行如下命令创建一个名称为 chapter6_1 的功能包:

```
$ cd ~/book_ws/src
$ ros2 pkg create --build-type ament_python chapter6_1
```

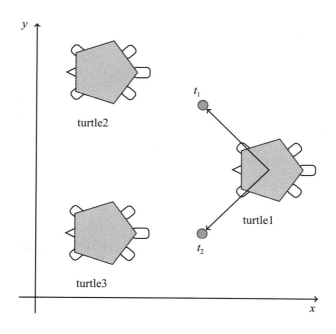

图 6.1 三只小海龟的领导–跟随示意图

然后构建这个包:

```
$ cd ~/book_ws
$ colcon build --packages-select chapter6_1
$ . install/setup.bash
```

通过如下命令创建一个实现小海龟状态广播的 Python 节点文件:

```
$ cd ~/book_ws/src/chapter6_1/chapter6_1
$ sudo nano turtle_pose_broadcaster.py
```

在 turtle_pose_broadcaster.py 文件中添加如下代码:

```python
#!/usr/bin/env python3
import rclpy
from rclpy.node import Node
from tf2_ros import TransformBroadcaster
from geometry_msgs.msg import TransformStamped
from tf_transformations import quaternion_from_euler
from turtlesim.msg import Pose

class PosePublisher(Node):
    def __init__(self):
        super().__init__('turtle_pose_publisher')
        self.declare_parameter('turtlename', 'turtle')
```

```python
13          self.turtlename = self.get_parameter('turtlename').get_parameter_value
                ().string_value
14          self.br = TransformBroadcaster(self)
15          self.subscription = self.create_subscription(
16              Pose,
17              f'/{self.turtlename}/pose',
18              self.publish_turtle_pose,
19              1)
20
21      def publish_turtle_pose(self, msg):
22          t = TransformStamped()
23          t.header.stamp = self.get_clock().now().to_msg()
24          t.header.frame_id = 'world'
25          t.child_frame_id = self.turtlename
26
27          t.transform.translation.x = msg.x
28          t.transform.translation.y = msg.y
29          t.transform.translation.z = 0.0
30
31          q = quaternion_from_euler(0, 0, msg.theta)
32          t.transform.rotation.x = q[0]
33          t.transform.rotation.y = q[1]
34          t.transform.rotation.z = q[2]
35          t.transform.rotation.w = q[3]
36
37          self.br.sendTransform(t)
38
39  def main():
40      rclpy.init()
41      node = PosePublisher()
42      try:
43          rclpy.spin(node)
44      except KeyboardInterrupt:
45          pass
46      rclpy.shutdown()
47
48  if __name__ == '__main__':
49      main()
```

下面解释上述代码的主要内容:

- 第 2~7 行导入需要用到的包,包括 TransformBroadcaster 用于对坐标变换进行广播,TransformStamped 是 geometry_msgs 的消息接口,用来实现对变换数据的定义，quaternion_from_euler 实现将欧拉角转换为四元数，Pose 是 turtlesim 的消息接口，用来读取小海龟的姿态信息；
- 第 9~37 行定义了一个节点类 PosePublisher() 用于发布小海龟的姿态，主要包括以下内容：
  - 第 11 行初始化节点名称为 turtle_pose_publisher；
  - 第 12 行用 declare_parameter 方法申明了名称为 turtlename 的参数，并将其默认值设置为 turtle，并在第 13 行通过 get_parameter 方法获取参数 turtlename 的值，这个值将在 launch 文件中给出，我们将在后续内容中介绍；
  - 第 14 行使用类 TransformBroadcaster() 初始化一个坐标变换的广播器，我们用它来发布需要的坐标变换；
  - 第 15~19 行通过 create_subscription 方法订阅小海龟的姿态信息，并通过回调函数 publish_turtle_pose() 在世界坐标系下进行发布；
  - 第 21~37 行定义了回调函数 publish_turtle_pose()，通过 TransformStamped 接口定义一个消息变量 t，并给其添加时间戳，定义其父坐标系名称为 world，子坐标系名称为参数 turtlename 的值，并将小海龟的位姿数据赋值给 t 的各个属性。

注 上述代码中导入的 turtlesim.msg 的 Pose 接口，其具体内容可以通过在终端窗口中运行如下命令查看[①]：

```
$ ros2 interface show turtlesim/msg/Pose
```

当运行上述命令时，终端窗口将显示如下内容：

```
float32 x
float32 y
float32 theta

float32 linear_velocity
float32 angular_velocity
```

该内容说明 turtlesim/msg/Pose 接口包含数据类型是 32 位浮点数的小海龟的坐标 x 和 y，角度 theta，以及线速度 linear_velocity 和角速度 angular_velocity。在代码的第 15 行通过定义一个订阅者 subscription 获取这些数据信息。此外，还可以通过在终端窗口中运行如下命令查看 geometry_msgs.msg 的 TransformStamped 接口[②]：

```
$ ros2 interface show geometry_msgs/msg/TransformStamped
```

当运行上述命令时，终端窗口除了显示一些说明文字，还显示了如下内容：

[①] turtlesim/msg/Pose 的内容也可以在如下网址查看：
http://docs.ros.org/en/api/turtlesim/html/msg/Pose.html.
[②] geometry_msgs/msg/TransformStamped 的内容也可以在如下网址查看：
https://docs.ros.org/en/api/geometry_msgs/html/msg/TransformStamped.html.

```
std_msgs/Header header
  builtin_interfaces/Time stamp
    int32 sec
    uint32 nanosec
  string frame_id

string child_frame_id

Transform transform
  Vector3 translation
    float64 x
    float64 y
    float64 z
  Quaternion rotation
    float64 x 0
    float64 y 0
    float64 z 0
    float64 w 1
```

该内容说明 geometry_msgs/msg/TransformStamped 接口包含信息 stamp、frame_id 和 child_frame_id, 关于位置的变换坐标 x, y 和 z, 以及关于姿态的旋转四元数 x, y, z 和 w. 我们在代码的 publish_turtle_pose() 回调函数中进行赋值并发布.

6.1.2 跟随目标状态发布

要实现如图6.1所示的领导-跟随任务, 三只小海龟需要保持一个相对位置固定的三角形队形: turtle2 与 turtle1 的相对位置是 $(-1,1)$, 即图中的 t_1 点, 而 turtle3 与 turtle1 的相对位置是 $(-1,-1)$, 即图中的 t_2 点. 因此, 当 turtle1 的绝对位置发生变化时, t_1 和 t_2 的绝对位置也会发生变化, 需要确定变化后的 t_1 和 t_2 的绝对位置, 这样两个跟随者 turtle2 和 turtle3 才能确定自己的移动目标点. 这里可以使用 ROS 2 将跟随者的移动目标点通过坐标变换发布出去, 以实现跟随者的移动.

在这一节中, 将通过静态广播方式发布小海龟跟随目标点的坐标位置, 以此确定跟随的小海龟需要到达的目标点.

在上述的 chapter6_1 的功能包中添加静态广播节点的代码, 通过如下命令创建一个实现静态广播的 Python 节点文件:

```
$ cd ~/book_ws/src/chapter6_1/chapter6_1
$ sudo nano follower_target_pose.py
```

在 follower_target_pose.py 文件中添加如下代码:

```python
#!/usr/bin/env python3
import rclpy
from rclpy.node import Node
from tf2_ros.static_transform_broadcaster import StaticTransformBroadcaster
from geometry_msgs.msg import TransformStamped
from tf_transformations import quaternion_from_euler

class FollowerTargetPublisher(Node):
    def __init__(self):
        super().__init__('follower_target_pose')
        self.declare_parameter('turtlename', 'turtle')
        self.declare_parameter('follower_target_x', 0.0)
        self.declare_parameter('follower_target_y', 0.0)
        self.turtlename = self.get_parameter('turtlename').get_parameter_value
            ().string_value
        self.followertargetx = self.get_parameter('follower_target_x').
            get_parameter_value().double_value
        self.followertargety = self.get_parameter('follower_target_y').
            get_parameter_value().double_value
        self.tf_publisher = StaticTransformBroadcaster(self)
        self.make_transform()

    def make_transform(self):
        st = TransformStamped()
        st.header.stamp = self.get_clock().now().to_msg()
        st.header.frame_id = 'turtle1'
        st.child_frame_id = f'{self.turtlename}_target'

        st.transform.translation.x = self.followertargetx
        st.transform.translation.y = self.followertargety
        st.transform.translation.z = 0.0

        quat = quaternion_from_euler(0.0, 0.0, 0.0)
        st.transform.rotation.x = quat[0]
        st.transform.rotation.y = quat[1]
        st.transform.rotation.z = quat[2]
        st.transform.rotation.w = quat[3]

        self.tf_publisher.sendTransform(st)

def main():
```

```
39      rclpy.init()
40      node = FollowerTargetPublisher()
41      try:
42          rclpy.spin(node)
43      except KeyboardInterrupt:
44          pass
45      rclpy.shutdown()
46
47  if __name__ == '__main__':
48      main()
```

下面解释上述代码的主要内容:

- 第 2~6 行导入需要用到的包,这些包与上一节的类似,不同的是因为跟随者的目标位置与 turtle1 保持相对静止,所以在代码的第 4 行导入了 StaticTransform-Broadcaster,用这个包实现静态坐标变换,并在代码的第 17 行进行了实例化;定义了变量 self.tf_publisher,用于存放静态坐标变换值;
- 代码的第 11~13 行申明了三个参数,分别是目标位置的名称,以及目标位置的 x 和 y 坐标,然后在代码的第 14~16 行获取这些参数的值 (参数的值将在 launch 文件中给出,后面再介绍);
- 代码的第 20~36 行给出了实现静态坐标变换函数 make_transform() 的内容,其中第 23 行定义了父坐标系是 turtle1,第 24 行定义了子坐标系是目标位置的名称.

6.1.3 跟随小海龟控制

在这一节中,将编写代码实现对跟随小海龟的控制,以达到领导-跟随效果的演示.主要通过减小跟随小海龟与目标点的位置差实现这一效果.

在上述 chapter6_1 的功能包中添加实现控制的代码,通过如下命令创建 Python 节点文件:

```
$ cd ~/book_ws/src/chapter6_1/chapter6_1
$ sudo nano follower_control.py
```

在 follower_control.py 文件中添加如下代码:

```
1  #!/usr/bin/env python3
2  import rclpy
3  from rclpy.node import Node
4  from geometry_msgs.msg import Twist
5  from tf2_ros import TransformException
6  from tf2_ros.buffer import Buffer
7  from tf2_ros.transform_listener import TransformListener
8  from turtlesim.srv import Spawn
```

```python
import math

class FrameListener(Node):
    def __init__(self):
        super().__init__('follower_control')
        self.declare_parameter('turtlename', 'turtle')
        self.turtlename = self.get_parameter('turtlename').get_parameter_value\
            ().string_value
        self.target_frame = f'{self.turtlename}_target'
        self.tf_buffer = Buffer()
        self.tf_listener = TransformListener(self.tf_buffer, self)
        self.spawner = self.create_client(Spawn, 'spawn')

        self.request = Spawn.Request()
        self.request.name = self.turtlename
        self.request.x = 4.0
        self.request.y = 2.0
        self.request.theta = 0.0
        self.result = self.spawner.call_async(self.request)

        self.publisher = self.create_publisher(Twist, f'/{self.turtlename}/\
            cmd_vel', 1)
        self.timer = self.create_timer(1.0, self.turtle_control)

    def turtle_control(self):
        from_frame_rel = self.target_frame
        to_frame_rel = self.turtlename
        try:
            now = rclpy.time.Time()
            trans = self.tf_buffer.lookup_transform(
                to_frame_rel,
                from_frame_rel,
                now)
        except TransformException as ex:
            pass

        msg = Twist()
        scale_forward_speed = 0.5
        msg.linear.x = scale_forward_speed * math.sqrt(
            trans.transform.translation.x ** 2 +
            trans.transform.translation.y ** 2)
```

```
49          scale_rotation_rate = 1.0
50          msg.angular.z = scale_rotation_rate * math.atan2(
51              trans.transform.translation.y,
52              trans.transform.translation.x)
53
54          self.publisher.publish(msg)
55
56  def main():
57      rclpy.init()
58      node = FrameListener()
59      try:
60          rclpy.spin(node)
61      except KeyboardInterrupt:
62          pass
63      rclpy.shutdown()
64
65  if __name__ == '__main__':
66      main()
```

下面解释上述代码的主要内容:

- 第 2~9 行加载了需要用到的包,其中第 4 行的 Twist 用于实现对小海龟的速度控制,第 7 行的 TransformListener 用于实现对坐标变换的监听,第 8 行的 Spawn 用于加载小海龟的仿真环境;
- 第 17~19 行定义了坐标变换的监听器,并启动了仿真环境客户端;
- 第 21~26 行初始化小海龟的状态,包括小海龟的名称,将初始位置设置成坐标 $(4,2)$,以及小海龟的头朝向与 x 轴一致,即在屏幕上朝向右方;
- 第 28 行初始化小海龟速度控制的发布者节点;
- 第 29 行调用计时器函数 turtle_control(),实现对小海龟的运动控制;
- 第 31~54 行给出了小海龟的控制函数,其大体思路是: 根据小海龟到目标点的距离差确定小海龟移动的线速度,根据小海龟与目标点的角度差确定小海龟移动的角速度. 控制函数的具体实现包括如下内容:
    - 第 32~41 行实现对小海龟目标位置相对于当前位置的实时监听,通过 lookup_transform 方法获得相对位置差;
    - 第 43~52 行通过实时监听到的坐标位置差计算小海龟的线速度和角速度,其中通过欧几里得范数确定线速度,通过反正切函数确定角速度;
    - 第 54 行将线速度和角速度发布给小海龟,控制其移动.

然后,打开 chapter6_1 目录下的 setup.py 文件,将 entry_points 域修改为如下形式:

```
1  entry_points={
2      'console_scripts': [
3          'turtle_pose_broadcaster = chapter6_1.turtle_pose_broadcaster:main',
4          'follower_target_pose = chapter6_1.follower_target_pose:main',
5          'follower_control = chapter6_1.follower_control:main',
6      ],
7  },
```

6.1.4 使用 launch 文件启动多个节点

因为打算启动三个小海龟实现领导–跟随任务,为了能够高效地启动多个节点,使用 launch 文件实现多个节点的启动. 在 chapter6_1 目录下新建一个名称为 launch 的目录,用于存放 launch 文件. 运行如下命令:

```
$ cd ~/book_ws/src/chapter6_1
$ mkdir launch
```

接下来, 在 launch 目录下新建一个名称为 turtle_leader_follower_demo.launch.py 的文件:

```
$ cd ~/book_ws/src/chapter6_1/launch
$ sudo nano turtle_leader_follower_demo.launch.py
```

并添加如下代码:

```
1   #!/usr/bin/env python3
2   from launch import LaunchDescription
3   from launch_ros.actions import Node
4
5   def generate_launch_description():
6       turtles = [
7           {'name': 'turtle1'},
8           {'name': 'turtle2', 'x_pose': -1.0, 'y_pose': 1.0},
9           {'name': 'turtle3', 'x_pose': -1.0, 'y_pose': -1.0}
10      ]
11
12      ld = LaunchDescription()
13
14      turtlesim_load_cmd = Node(
15          package = 'turtlesim',
16          executable = 'turtlesim_node',
17          name = 'sim'
```

```
18              )
19          ld.add_action(turtlesim_load_cmd)
20
21          for i in range(0, 3):
22              turtle_pose_broadcaster_cmd = Node(
23                  package = 'chapter6_1',
24                  executable = 'turtle_pose_broadcaster',
25                  name = turtles[i].get('name') + '_node',
26                  parameters = [
27                      {'turtlename': turtles[i].get('name')}
28                  ]
29              )
30              ld.add_action(turtle_pose_broadcaster_cmd)
31
32          for i in range(1, 3):
33              follower_target_pose_cmd = Node(
34                  package = 'chapter6_1',
35                  executable = 'follower_target_pose',
36                  name = turtles[i].get('name') + '_target_node',
37                  parameters = [
38                      {'turtlename': turtles[i].get('name')},
39                      {'follower_target_x': turtles[i].get('x_pose')},
40                      {'follower_target_y': turtles[i].get('y_pose')}
41                  ]
42              )
43              ld.add_action(follower_target_pose_cmd)
44
45              follower_control_cmd = Node(
46                  package = 'chapter6_1',
47                  executable = 'follower_control',
48                  name = turtles[i].get('name') + '_control_node',
49                  parameters = [
50                      {'turtlename': turtles[i].get('name')}
51                  ]
52              )
53              ld.add_action(follower_control_cmd)
54
55          return ld
```

下面解释上述代码的主要内容:
- 代码的第 2~3 行加载了运行此 launch 文件需要用到的包;
- 第 6~10 行定义了三只小海龟的参数,包括名称和两个跟随者相对领导者的坐标

位置差;
- 第 12 行定义了变量 ld 用于存储后面将要加载的节点文件;
- 第 14~19 行定义了小海龟的仿真环境并将其添加到变量 ld 中;
- 第 21~30 行定义了小海龟的位姿发布节点, 使用了我们前面编写的 turtle_pose_broadcaster.py 文件, 并将其添加到变量 ld 中;
- 第 32~53 行定义了两只跟随小海龟的目标位置节点和速度控制节点, 使用了前面编写的 follower_target_pose.py 和 follower_control.py 文件, 并将其添加到变量 ld 中.

重新打开 chapter6_1 目录下的 setup.py 文件中修改 data_files 域为如下形式:

```
data_files=[
    ...
    (os.path.join('share', package_name, 'launch'), glob(os.path.join('launch', '*.launch.py'))),
],
```

并在文件头部添加如下代码:

```
import os
from glob import glob
```

重新构建 chapter6_1 功能包, 运行如下命令:

```
$ cd ~/book_ws
$ colcon build --packages-select chapter6_1
```

打开两个终端, 分别输入如下两条命令进行测试:

```
$ ros2 launch chapter6_1 turtle_leader_follower_demo.launch.py
$ ros2 run turtlesim turtle_teleop_key
```

在后一条命令对应的终端中, 可以通过键盘的方向键控制小海龟移动. 图6.2显示了初始状态和通过键盘的方向键控制后的移动效果.

此外, 可以通过如下命令显示通过 ROS 2 系统的 tf2 正在广播的坐标变换树 (TF Tree) 的图像:

```
$ ros2 run tf2_tools view_frames
```

如图6.3所示, 世界坐标系 world 是根节点, 三只小海龟都是 world 的子节点, 两个跟随者都是 turtle1 的子节点. 这些节点构成了一棵连通的树形结构, 因此可以监测任意两个节点之间的坐标变换. 在前面的程序中, 分别监测了 turtle2 与 turtle2_target 和 turtle3 与 turtle3_target 之间的坐标变换, 以实现跟随移动效果.

也可以通过如下命令可视化节点和话题的关系图 (图6.4):

```
$ rqt_graph
```

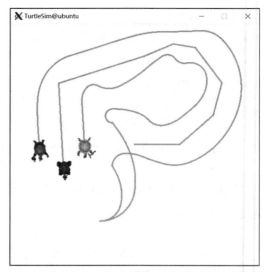

初始状态　　　　　　　　　　　通过键盘的方向键控制 turtle1 移动后的效果

图 6.2　三只小海龟的领导–跟随效果

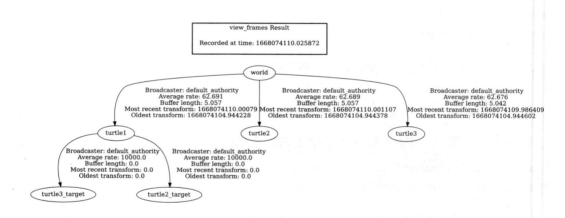

图 6.3　实现三只小海龟领导–跟随任务的坐标变换树

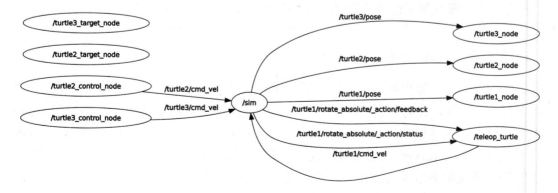

图 6.4　实现三只小海龟领导–跟随任务的节点和话题关系图

6.2 在 Gazebo 仿真中实现 TurtleBot 3 的领导–跟随任务

上一节介绍的 turtlesim 仿真环境下实现小海龟的领导–跟随任务，虽然能够达到一定的仿真效果，但是和真实的机器人环境还是相差较大. 因此，有必要使用具有实际物理特征的机器人进行仿真. 在这一节中，将进一步在 Gazebo 仿真中实现三个 TurtleBot 3 机器人的领导–跟随任务，其实现原理和小海龟的程序相似.

6.2.1 TurtleBot 3 状态发布

首先，介绍如何将 TurtleBot 3 机器人的状态在世界坐标系下发布出去，以用于实现后续的领导–跟随任务.

执行如下命令创建一个名称为 chapter6_2 的功能包:

```
$ cd ~/book_ws/src
$ ros2 pkg create --build-type ament_python chapter6_2
```

然后构建这个包:

```
$ cd ~/book_ws
$ colcon build --packages-select chapter6_2
$ . install/setup.bash
```

通过如下命令创建一个实现 TurtleBot 3 机器人状态广播的 Python 节点文件:

```
$ cd ~/book_ws/src/chapter6_2/chapter6_2
$ sudo nano tb3_odom_broadcaster.py
```

在 tb3_odom_broadcaster.py 文件中添加如下代码:

```python
#!/usr/bin/env python3
import rclpy
from rclpy.node import Node
from tf2_ros import TransformBroadcaster
from geometry_msgs.msg import TransformStamped
from nav_msgs.msg import Odometry

class OdomPublisher(Node):
    def __init__(self):
        super().__init__('tb3_odom_publisher')
        self.declare_parameter('tb3name', 'turtlebot')
        self.tb3name = self.get_parameter('tb3name').get_parameter_value().string_value
        self.br = TransformBroadcaster(self)
```

```
14            self.subscription = self.create_subscription(
15                Odometry,
16                f'/{self.tb3name}/odom',
17                self.publish_tb3_odom,
18                1)
19
20        def publish_tb3_odom(self, msg):
21            t = TransformStamped()
22            t.header.stamp = self.get_clock().now().to_msg()
23            t.header.frame_id = 'world'
24            t.child_frame_id = f'/{self.tb3name}/odom'
25
26            t.transform.translation.x = msg.pose.pose.position.x
27            t.transform.translation.y = msg.pose.pose.position.y
28            t.transform.translation.z = 0.0
29
30            t.transform.rotation.x = msg.pose.pose.orientation.x
31            t.transform.rotation.y = msg.pose.pose.orientation.y
32            t.transform.rotation.z = msg.pose.pose.orientation.z
33            t.transform.rotation.w = msg.pose.pose.orientation.w
34
35            self.br.sendTransform(t)
36
37    def main():
38        rclpy.init()
39        node = OdomPublisher()
40        try:
41            rclpy.spin(node)
42        except KeyboardInterrupt:
43            pass
44        rclpy.shutdown()
45
46    if __name__ == '__main__':
47        main()
```

此代码与小海龟仿真中的相应代码类似，因此仅对其中不同的地方进行解释：

- 代码的第 6 行从 ROS 2 的导航包中加载了 Odometry 消息类型，使用这个接口可以获得机器人在空间中的位姿；
- 第 14~18 行订阅了机器人的里程计数据，并通过回调函数 publish_tb3_odom() 完成里程计数据相对世界坐标系 world 的坐标变换，这一操作相当于实现了对机器人的局部坐标系和全局坐标系的坐标变换；

- 第 20~35 行定义了实现上述坐标变换的回调函数. 对每个机器人而言, 该回调函数将机器人当前的里程计数据相对于 world 坐标系进行了坐标变换, 这样所有机器人都有了一个共同的全局坐标系.

6.2.2 跟随目标状态发布

接下来, 通过静态广播方式发布跟随机器人的目标状态, 以此确定跟随机器人需要到达的目标位置.

在上述的 chapter6_2 功能包中添加静态广播节点的代码, 通过如下命令创建一个实现静态广播的 Python 节点文件:

```
$ cd ~/book_ws/src/chapter6_2/chapter6_2
$ sudo nano follower_target_pose.py
```

在 follower_target_pose.py 文件中添加如下代码:

```
1  #!/usr/bin/env python3
2  import rclpy
3  from rclpy.node import Node
4  from tf2_ros.static_transform_broadcaster import StaticTransformBroadcaster
5  from geometry_msgs.msg import TransformStamped
6  from tf_transformations import quaternion_from_euler
7
8  class FollowerTargetPublisher(Node):
9    def __init__(self):
10     super().__init__('follower_target_pose')
11     self.declare_parameter('tb3name', 'turtlebot')
12     self.declare_parameter('follower_target_x', 0.0)
13     self.declare_parameter('follower_target_y', 0.0)
14     self.tb3name = self.get_parameter('tb3name').get_parameter_value().
           string_value
15     self.followertargetx = self.get_parameter('follower_target_x').
           get_parameter_value().double_value
16     self.followertargety = self.get_parameter('follower_target_y').
           get_parameter_value().double_value
17     self.tf_publisher = StaticTransformBroadcaster(self)
18     self.make_transform()
19
20   def make_transform(self):
21     st = TransformStamped()
22     st.header.stamp = self.get_clock().now().to_msg()
23     st.header.frame_id = '/robot1/odom'
24     st.child_frame_id = f'/{self.tb3name}/target'
```

```
25
26            st.transform.translation.x = self.followertargetx
27            st.transform.translation.y = self.followertargety
28            st.transform.translation.z = 0.0
29
30            quat = quaternion_from_euler(0.0, 0.0, 0.0)
31            st.transform.rotation.x = quat[0]
32            st.transform.rotation.y = quat[1]
33            st.transform.rotation.z = quat[2]
34            st.transform.rotation.w = quat[3]
35
36            self.tf_publisher.sendTransform(st)
37
38   def main():
39       rclpy.init()
40       node = FollowerTargetPublisher()
41       try:
42           rclpy.spin(node)
43       except KeyboardInterrupt:
44           pass
45       rclpy.shutdown()
46
47   if __name__ == '__main__':
48       main()
```

这个代码和第6.1.2节的类似，不同的是这里在第 23 行使用机器人的里程计数据代替了小海龟的状态数据. 其他代码基本相同，这里就不再详细解释了.

6.2.3 跟随机器人控制

在这一节中，将实现对跟随机器人的控制，以实现领导−跟随效果的演示. 通过减小跟随机器人和目标位置的差异，实现这一效果.

在上述的 chapter6_2 功能包中添加实现机器人控制的代码，通过如下命令创建一个实现实时机器人控制的 Python 节点文件:

```
$ cd ~/book_ws/src/chapter6_2/chapter6_2
$ sudo nano follower_control.py
```

在 follower_control.py 文件中添加如下代码:

```
1   #!/usr/bin/env python3
2   import rclpy
3   from rclpy.node import Node
```

```python
from geometry_msgs.msg import Twist
from tf2_ros import TransformException
from tf2_ros.buffer import Buffer
from tf2_ros.transform_listener import TransformListener
import math

class FrameListener(Node):
    def __init__(self):
        super().__init__('follower_control')
        self.declare_parameter('tb3name', 'turtlebot')
        self.tb3name = self.get_parameter('tb3name').get_parameter_value().\
            string_value
        self.tf_buffer = Buffer()
        self.tf_listener = TransformListener(self.tf_buffer, self)
        self.publisher = self.create_publisher(Twist, f'/{self.tb3name}/\
            cmd_vel', 1)
        self.timer = self.create_timer(0.1, self.tb3_control)

    def tb3_control(self):
        from_frame_rel = f'{self.tb3name}/target'
        to_frame_rel = f'{self.tb3name}/odom'
        try:
            now = rclpy.time.Time()
            trans = self.tf_buffer.lookup_transform(
                to_frame_rel,
                from_frame_rel,
                now)
        except TransformException as ex:
            return

        msg = Twist()
        scale_forward_speed = 0.5
        msg.linear.x = scale_forward_speed * math.sqrt(
            trans.transform.translation.x ** 2 +
            trans.transform.translation.y ** 2)

        scale_rotation_rate = 1
        msg.angular.z = scale_rotation_rate * math.atan2(
            trans.transform.translation.y,
            trans.transform.translation.x)

        self.publisher.publish(msg)
```

```
44
45  def main():
46      rclpy.init()
47      node = FrameListener()
48      try:
49          rclpy.spin(node)
50      except KeyboardInterrupt:
51          pass
52      rclpy.shutdown()
53
54  if __name__ == '__main__':
55      main()
```

这个代码和第6.1.3节的类似,不同的是这里在第 22 行使用机器人的里程计数据代替了小海龟的状态数据. 此外,这里没有加载机器人仿真环境的代码,对机器人仿真环境的加载将在 launch 文件里实现,这将在下一节具体介绍.

接下来,打开 chapter6_1 目录下的 setup.py 文件, 将 entry_points 域修改为如下形式:

```
entry_points={
    'console_scripts': [
        'turtle_pose_broadcaster = chapter6_2.turtle_pose_broadcaster:main',
        'follower_target_pose = chapter6_2.follower_target_pose:main',
        'follower_control = chapter6_2.follower_control:main',
    ],
},
```

重新构建 chapter6_2 功能包,运行如下命令:

```
$ cd ~/book_ws
$ colcon build --packages-select chapter6_2
```

6.2.4 使用 launch 文件启动多个节点

因为打算启动三个 TurtleBot 3 机器人实现领导–跟随任务,为了能够高效地启动多个节点,同样使用 launch 文件实现多个节点的启动. 在 chapter6_2 目录下新建一个 launch 目录,用以存放 launch 文件,运行如下命令:

```
$ cd ~/book_ws/src/chapter6_2
$ mkdir launch
```

接下来,在 launch 目录下新建一个名称为 tb3_leader_follower_demo.launch.py 的文件:

```
$ cd ~/book_ws/src/chapter6_2/launch
$ sudo nano tb3_leader_follower_demo.launch.py
```

并添加如下代码:

```python
#!/usr/bin/env python3
import os
from ament_index_python.packages import get_package_share_directory
from launch import LaunchDescription
from launch.actions import IncludeLaunchDescription
from launch.launch_description_sources import PythonLaunchDescriptionSource
from launch_ros.actions import Node

def generate_launch_description():
    TURTLEBOT3_MODEL = os.environ['TURTLEBOT3_MODEL']
    model_folder = 'turtlebot3_' + TURTLEBOT3_MODEL
    sdf_path = os.path.join(
        get_package_share_directory('turtlebot3_gazebo'),
        'models',
        model_folder,
        'model.sdf'
        )

    urdf_file_name = 'turtlebot3_' + TURTLEBOT3_MODEL + '.urdf'
    urdf_path = os.path.join(
        get_package_share_directory('turtlebot3_description'),
        'urdf',
        urdf_file_name)
    with open(urdf_path, 'r') as infp:
        robot_description = infp.read()

    default_world_path = os.path.join(
        get_package_share_directory('turtlebot3_gazebo'),
        'worlds',
        'empty_world.world'
        )

    robots = [
        {'name': 'robot1', 'x_pose': '0.0', 'y_pose': '0.0', 'z_pose': '0.01'
            },
        {'name': 'robot2', 'x_pose': '0.0', 'y_pose': '1.0', 'z_pose': '0.01',
            'x_target': -1.0, 'y_target': 1.0},
```

```python
        {'name': 'robot3', 'x_pose': '0.0', 'y_pose': '-1.0', 'z_pose': '0.01'
            , 'x_target': -1.0, 'y_target': -1.0}
    ]

    gazebo_load_cmd = IncludeLaunchDescription(
        PythonLaunchDescriptionSource(
            os.path.join(get_package_share_directory('gazebo_ros'), 'launch',
                'gazebo.launch.py')
        ),
        launch_arguments = {'world': default_world_path}.items()
    )

    ld = LaunchDescription()
    ld.add_action(gazebo_load_cmd)

    for i in range(0, 3):
        spawn_robot_cmd = Node(
            package = 'gazebo_ros',
            executable = 'spawn_entity.py',
            name = robots[i].get('name')+'_spawn_node',
            output = 'screen',
            arguments = [
                '-entity', robots[i].get('name')+'_entity',
                '-robot_namespace', robots[i].get('name'),
                '-file', sdf_path,
                '-x', robots[i].get('x_pose'),
                '-y', robots[i].get('y_pose'),
                '-z', robots[i].get('z_pose'),
            ],
        )
        ld.add_action(spawn_robot_cmd)

        robot_state_publisher_cmd = Node(
            package = 'robot_state_publisher',
            executable = 'robot_state_publisher',
            namespace = robots[i].get('name'),
            name = robots[i].get('name')+'_state_node',
            output = 'screen',
            parameters = [{
                'use_sim_time': True,
                'robot_description': robot_description,
                'publish_frequency': 50.0
```

```
76              }],
77          )
78          ld.add_action(robot_state_publisher_cmd)
79
80          tb3_odom_broadcaster_cmd = Node(
81              package = 'chapter6_2',
82              executable = 'tb3_odom_broadcaster',
83              name = robots[i].get('name')+'_odom_br',
84              parameters = [{
85                  'tb3name': robots[i].get('name'),
86              }],
87          )
88          ld.add_action(tb3_odom_broadcaster_cmd)
89
90      for i in range(1, 3):
91          tb3_follower_target_cmd = Node(
92              package = 'chapter6_2',
93              executable = 'follower_target_pose',
94              name = robots[i].get('name')+'_target_node',
95              parameters = [{
96                  'tb3name': robots[i].get('name'),
97                  'follower_target_x': robots[i].get('x_target'),
98                  'follower_target_y': robots[i].get('y_target')
99              }],
100         )
101         ld.add_action(tb3_follower_target_cmd)
102
103         tb3_follower_control_cmd = Node(
104             package = 'chapter6_2',
105             executable = 'follower_control',
106             name = robots[i].get('name')+'_follower_node',
107             parameters = [{
108                 'tb3name': robots[i].get('name')
109             }],
110         )
111         ld.add_action(tb3_follower_control_cmd)
112
113     return ld
```

下面解释上述代码的主要内容:
- 第 2~7 行加载了运行此 launch 文件需要用到的包;
- 第 10~17 行定义了 TurtleBot 3 模型的 sdf 文件路径;

- 第 19~25 行定义了 TurtleBot 3 模型的 urdf 文件路径；
- 第 27~31 行定义了 Gazebo 仿真环境需要加载的地图路径，这里使用空地图；
- 第 33~37 行定义了变量 robots 用于存储三个机器人的基本信息，包括名称和初始坐标位置，以及两个跟随者相对领导者的坐标位置差；
- 第 39~44 行定义了加载的 Gazebo 仿真环境；
- 第 46~47 行定义了变量 ld 用于存储节点变量，并加载了 Gazebo 仿真环境；
- 第 50~64 行通过运行 gazebo_ros 包的 spawn_entity.py 节点文件加载机器人模型并初始化；
- 第 66~78 行通过运行 robot_state_publisher 包的 robot_state_publisher 节点文件发布机器人传感器的状态信息；
- 第 80~88 行通过运行 tb3_odom_broadcaster 节点文件发布机器人的里程计相对全局坐标系的坐标变换；
- 同样地，第 90~111 行通过运行节点文件发布跟随机器人的目标位置相对全局坐标系的坐标变化，以及实现对跟随机器人的速度控制。

接下来，重新打开 chapter6_2 目录下的 setup.py 文件，将 data_files 域修改为如下形式：

```
1  data_files=[
2      ...
3      (os.path.join('share', package_name, 'launch'), glob(os.path.join('launch', '*.launch.py'))),
4  ],
```

并在文件头部添加如下代码：

```
1  import os
2  from glob import glob
```

重新构建 chapter6_2 功能包，运行如下命令：

```
$ cd ~/book_ws
$ colcon build --packages-select chapter6_2
```

打开两个终端，分别输入如下两条命令进行测试：

```
$ ros2 launch chapter6_2 tb3_leader_follower_demo.launch.py
$ ros2 topic pub --once /robot1/cmd_vel geometry_msgs/msg/Twist "{linear: {x: 0.1, y: 0.0, z: 0.0}, angular: {x: 0.0, y: 0.0, z: 0.03}}"
```

图6.5给出了在 Gazebo 仿真中实现三个 TurtleBot 3 机器人的领导–跟随演示效果。可以看到，领导机器人 robot1 沿弧线往前行进，两个跟随机器人 robot2 和 robot3 分别在 robot1 的左右后方前进，并形成三角形队形。

可以通过如下命令可视化节点和话题的关系图：

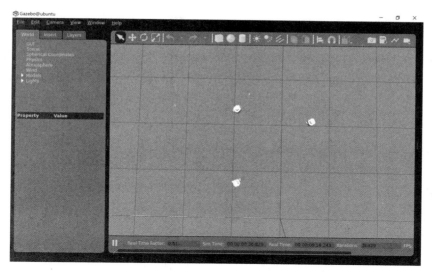

图 6.5　在 Gazebo 仿真中三个 TurtleBot 3 机器人的领导–跟随演示效果

```
$ rqt_graph
```

如图6.6所示，可以看到节点和话题的关系图比小海龟仿真的图6.4要复杂得多．这是由于 TurtleBot 3 机器人多了许多传感器，因此会多出一些关于传感器的节点和话题信息．

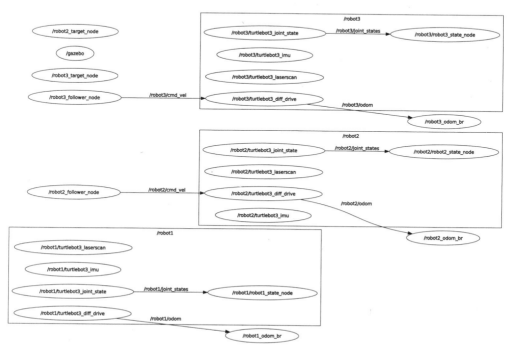

图 6.6　实现三个 TurtleBot 3 机器人领导–跟随任务的节点和话题关系图

练 习

1. (网络信息传播与同步)假设有 10 个节点构成一个无向连通图, 每个节点至少有两个邻居. 每个节点随机发布自己的状态(整数 0 或者 1), 并且同时订阅邻居的状态. 对于任意一个节点, 如果它订阅的所有邻居的状态相同(即都是 0 或者都是 1), 则这个节点只发布状态值 1. 试编写代码并利用 ROS 2 的通信机制实现上述描述.

2. 试修改第6.2节的代码, 实现五个 TurtleBot 3 机器人的领导–跟随任务. 队形要求: robot2~robot5 作为跟随者保持正方形队形, robot1 作为领导者位于正方形队形的中心位置.

第 7 章 基于集中式和分布式优化算法的 ROS 2 仿真

在这一章中,将基于前述章节的分布式优化算法,在 ROS 2 的 Gazebo 仿真中实现多机器人的队形切换.要实现多机器的队形切换,需要分两步来完成:第一步是实现单个机器人按照指定坐标点的定点移动,第二步是编写优化算法实现多个机器人按照最优目标位置的定点移动.

7.1 实现单个机器人的定点移动

在这一节中,将实现单个 TurtleBot 3 机器人的定点移动.执行如下命令创建一个名称为 chapter7_1 的功能包:

```
$ cd ~/book_ws/src
$ ros2 pkg create --build-type ament_python chapter7_1
```

然后构建这个包:

```
$ cd ~/book_ws
$ colcon build --packages-select chapter7_1
$ . install/setup.bash
```

7.1.1 编写定点移动代码

通过如下命令创建一个实现定点移动的 Python 节点文件:

```
$ cd ~/book_ws/src/chapter7_1/chapter7_1
$ sudo nano move_to_goal.py
```

在 move_to_goal.py 文件中添加如下代码:

```
1  #!/usr/bin/env python3
2  import rclpy
3  from rclpy.node import Node
```

```python
from nav_msgs.msg import Odometry
from geometry_msgs.msg import Twist
from tf_transformations import euler_from_quaternion
import math

def quat_to_angle(quat):
    rot = euler_from_quaternion((quat.x, quat.y, quat.z, quat.w))
    return rot

def normalize_angle(angle):
    res = angle
    while res > math.pi:
        res -= 2.0*math.pi
    while res < -math.pi:
        res += 2.0*math.pi
    return res

class RobotControl(Node):
    def __init__(self):
        super().__init__('robot_control')
        self.declare_parameter('tb3name', 'turtlebot')
        self.tb3name = self.get_parameter('tb3name').get_parameter_value().
            string_value
        self.state_x = 0.0
        self.state_y = 0.0
        self.theta = 0.0

        self.odom_sub = self.create_subscription(
            Odometry,
            f'/{self.tb3name}/odom',
            self.current_state,
            1)

        self.robot_vel = self.create_publisher(
            Twist,
            f'/{self.tb3name}/cmd_vel',
            1)

    def current_state(self, data):
        self.state_x = data.pose.pose.position.x
        self.state_y = data.pose.pose.position.y
        (roll, pitch, self.theta) = quat_to_angle(data.pose.pose.orientation)
```

```python
    def robot_control(self, goal):
        goal_x = goal[0]
        goal_y = goal[1]
        is_arrival = False
        tolerance = 0.02

        trans = (goal_x - self.state_x, goal_y - self.state_y)
        linear = math.sqrt(trans[0] ** 2 + trans[1] ** 2)
        angular = normalize_angle(math.atan2(trans[1], trans[0]) - self.theta)
        msg = Twist()

        if linear > tolerance:
            if abs(angular) > math.pi/8:
                msg.linear.x = min(linear*0.5, 0.22)
                msg.angular.z = max(min(angular*0.8, 2.84), -2.84)
            else:
                msg.linear.x = min(linear*1.0, 0.22)
                msg.angular.z = max(min(angular*0.1, 2.84), -2.84)
        else:
            msg.linear.x = 0.0
            msg.angular.z = 0.0
            is_arrival = True
        self.robot_vel.publish(msg)
        return is_arrival

def main():
    rclpy.init()
    goals = [[1.0, 1.0], [1.0, 0.0], [0.0, 0.0]]

    for i in range(0, len(goals)):
        goal = goals[i]

        node = RobotControl()
        node.get_logger().info('\n 当前机器人的移动目标: ' + str(goal))
        while rclpy.ok():
            try:
                rclpy.spin_once(node)
                is_arrival = node.robot_control(goal)
            except KeyboardInterrupt:
                pass
            if is_arrival == True:
```

```
87                    break
88         rclpy.shutdown()
89
90  if __name__ == '__main__':
91      main()
```

下面解释上述代码的主要内容:

- 第 2~7 行加载了需要用到的包,其中第 4 行 Odometry 是关于机器人里程计的消息接口,第 5 行的 Twist 用于实现对机器人进行速度控制的消息接口;
- 第 9~19 行定义了两个函数,其中函数 quat_to_angle() 用于将表示机器人姿态的四元数转化为欧拉角,函数 normalize_angle() 将角度进行标准化,即转化到区间 $[-\pi, \pi]$;
- 第 21~69 行定义了类 RobotControl(),主要包括以下内容:
  - 第 26~28 行是将机器人的位姿信息进行初始化. 由于这里使用的是无人车,因此只需要考虑无人车在二维平面上的坐标变量 state_x 和 state_y, 以及表示无人车朝向的角度变量 theta;
  - 第 30~34 行用于订阅机器人的里程计信息,并通过回调函数 current_state() 将里程计信息赋值给全局变量 state_x, state_y 和 theta;
  - 第 36~39 行定义了用于机器人速度控制的发布者,其发布的速度信息在函数 robot_control() 中获得;
  - 第 46~69 行给出了对机器人进行速度控制的函数,其计算原理是: 根据机器人当前的坐标位置和朝向,计算机器人与目标位置的距离差和角度差,并将其赋值给变量 msg, 最后将线速度和角速度通过 Twist 接口发布给机器人.
- 第 71~88 行的主程序设置了三个目标位置,分别是 (1,1), (1,0) 和 (0,0). 第 83 行对机器人进行速度控制的同时返回变量 is_arrival 的值,用于判断机器人是否到达当前的目标位置. 如果到达,则跳出循环,机器人向下一个目标位置移动.

打开 chapter7_1 目录下的 setup.py 文件,将 entry_points 域修改为如下形式:

```
1  entry_points={
2      'console_scripts': [
3          'move_to_goal = chapter7_1.move_to_goal:main',
4      ],
5  },
```

7.1.2 编写 launch 文件测试定点移动效果

在这一节中,编写一个 launch 文件测试上述定点移动代码的效果. 在 chapter7_1 目录下新建一个 launch 目录,用于存放 launch 文件,运行如下命令:

```
$ cd ~/book_ws/src/chapter7_1
$ mkdir launch
```

接下来，在launch目录下新建一个名称为tb3_move_to_goal_demo.launch.py的文件：

```
$ cd ~/book_ws/src/chapter7_1/launch
$ sudo nano tb3_move_to_goal_demo.launch.py
```

并添加如下代码：

```python
#!/usr/bin/env python3
import os
from ament_index_python.packages import get_package_share_directory
from launch import LaunchDescription
from launch.actions import IncludeLaunchDescription
from launch.launch_description_sources import PythonLaunchDescriptionSource
from launch_ros.actions import Node

def generate_launch_description():
    TURTLEBOT3_MODEL = os.environ['TURTLEBOT3_MODEL']
    model_folder = 'turtlebot3_' + TURTLEBOT3_MODEL
    sdf_path = os.path.join(
        get_package_share_directory('turtlebot3_gazebo'),
        'models',
        model_folder,
        'model.sdf'
        )

    urdf_file_name = 'turtlebot3_' + TURTLEBOT3_MODEL + '.urdf'
    urdf_path = os.path.join(
        get_package_share_directory('turtlebot3_description'),
        'urdf',
        urdf_file_name)
    with open(urdf_path, 'r') as infp:
        robot_description = infp.read()

    default_world_path = os.path.join(
        get_package_share_directory('turtlebot3_gazebo'),
        'worlds',
        'empty_world.world'
        )

```

```python
33      robot = {'name': 'robot', 'x_pose': '0.0', 'y_pose': '0.0', 'z_pose': '
            0.01'}
34
35      gazebo_load_cmd = IncludeLaunchDescription(
36          PythonLaunchDescriptionSource(
37              os.path.join(get_package_share_directory('gazebo_ros'), 'launch',
                    'gazebo.launch.py')
38          ),
39          launch_arguments = {'world': default_world_path}.items()
40      )
41
42      ld = LaunchDescription()
43      ld.add_action(gazebo_load_cmd)
44
45      spawn_robot_cmd = Node(
46          package = 'gazebo_ros',
47          executable = 'spawn_entity.py',
48          name = robot.get('name')+'_spawn_node',
49          output = 'screen',
50          arguments = [
51              '-entity', robot.get('name')+'_entity',
52              '-robot_namespace', robot.get('name'),
53              '-file', sdf_path,
54              '-x', robot.get('x_pose'),
55              '-y', robot.get('y_pose'),
56              '-z', robot.get('z_pose'),
57          ],
58      )
59      ld.add_action(spawn_robot_cmd)
60
61      robot_state_publisher_cmd = Node(
62          package = 'robot_state_publisher',
63          executable = 'robot_state_publisher',
64          namespace = robot.get('name'),
65          name = robot.get('name')+'_state_node',
66          output = 'screen',
67          parameters = [{
68              'use_sim_time': True,
69              'robot_description': robot_description,
70              'publish_frequency': 50.0
71          }],
72      )
```

```
73      ld.add_action(robot_state_publisher_cmd)
74
75      tb3_move_to_goal_cmd = Node(
76          package = 'chapter7_1',
77          executable = 'move_to_goal',
78          name = robot.get('name')+'_move_to_goal_node',
79          parameters = [{
80              'tb3name': robot.get('name')
81              }],
82          )
83      ld.add_action(tb3_move_to_goal_cmd)
84
85      return ld
```

上述代码与第6.2.4节的 launch 文件类似, 不同的地方是这里的第 75~83 加载了前面定义的机器人定点移动的节点文件.

重新打开 chapter7_1 目录下的 setup.py 文件, 将 data_files 域修改为如下形式:

```
1  data_files=[
2      ...
3      (os.path.join('share', package_name, 'launch'), glob(os.path.join('launch
       ', '*.launch.py'))),
4  ],
```

并在文件头部添加如下代码:

```
1  import os
2  from glob import glob
```

接下来, 重新构建 chapter7_1 功能包, 运行如下命令:

```
$ cd ~/book_ws
$ colcon build --packages-select chapter7_1
```

打开终端, 输入如下命令进行测试:

```
$ ros2 launch chapter7_1 tb3_move_to_goal_demo.launch.py
```

图7.1给出了在 Gazebo 中的仿真结果, 可以看到机器人从 (0,0) 点出发, 首先移动到 (1,1) 点, 再到 (1,0) 点, 最后回到 (0,0) 点.

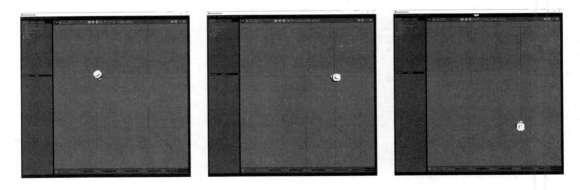

图 7.1　在 Gazebo 仿真中 TurtleBot 3 机器人的定点移动

7.2　多个机器人队形切换——集中式优化算法

在这一节中，将在 Gazebo 仿真中实现三个 TurtleBot 3 机器人的队形切换任务，队形由 robot2/robot3 与 robot1 的相对位置确定，即考虑第2章中无参考中心的编队优化模型.

如图7.2所示，我们考虑三个队形的切换问题，即机器人从直线队形出发，首先变换成三角形队形(队形1)，然后变换成直线队形(队形2)，最后再变换成三角形队形(队形3). 第一个队形中，robot1 与 robot2 和 robot3 的相对位置分别是$(1,-1)$和$(1,1)$；第二个队形中，robot1 与 robot2 和 robot3 的相对位置分别是$(0,-1)$和$(0,1)$；第三个队形中，robot1 与 robot2 和 robot3 的相对位置分别是$(0,1)$和$(-1,0)$.

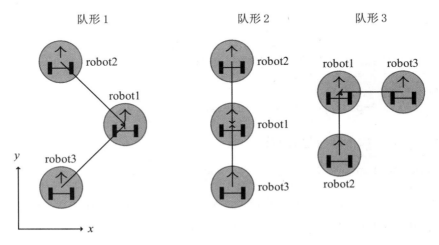

图 7.2　本节中考虑的三个机器人形成的三个队形

为了实现上述的队形切换，首先考虑集中式优化算法如何解决此类问题. 设计框架如下所述：有三个机器人节点和一个中心节点，三个机器人节点订阅中心节点发布的最优队形话题并移动到自己的最优目标位置,中心节点订阅三个机器人的里程计话题并计算最优队形,然后将最优队形发布给每个机器人. 此设计框架如图7.3所示,用一句话概括,即是集中式计算分布式控制.

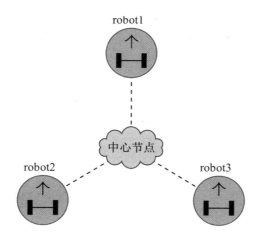

图 7.3　集中式计算分布式控制框架

将编写代码实现上述队形切换任务,算法流程如图7.4所示. 从该流程图中可以看出,主要包括两个程序文件: 一个是 optimal_formation_publisher.py 文件,用于中心节点订阅机器人的里程计话题,计算最优队形和发布最优队形;另一个是 tb3_move_to_optimal.py 文件,用于机器人订阅最优队形和控制机器人移动到目标位置. 接下来,将具体编写这两个程序文件.

执行如下命令创建一个名称为 chapter7_2 的功能包:

```
$ cd ~/book_ws/src
$ ros2 pkg create --build-type ament_python chapter7_2
```

然后构建这个包:

```
$ cd ~/book_ws
$ colcon build --packages-select chapter7_2
$ . install/setup.bash
```

7.2.1　计算最优队形

根据第2章无参考中心的编队优化模型所对应的优化算法,编写计算最优队形的代码. 在 chapter7_2 功能包中创建一个计算并发布最优队形的 Python 节点文件:

```
$ cd ~/book_ws/src/chapter7_2/chapter7_2
$ sudo nano optimal_formation_publisher.py
```

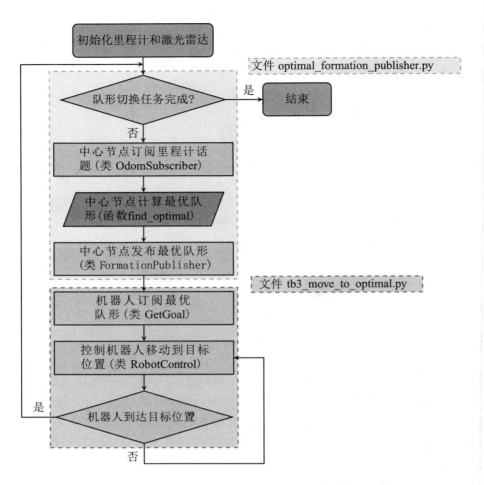

图 7.4　实现队形切换的集中式算法流程图

在 optimal_formation_publisher.py 文件中添加如下代码：

```python
#!/usr/bin/env python3
import rclpy
from rclpy.node import Node
from nav_msgs.msg import Odometry
from std_msgs.msg import Float32MultiArray
import numpy as np

def proj(x, lb, ub):
    y = x.copy()
    y[x>ub] = ub
    y[x<lb] = lb
    return y

def find_optimal(Delta, p0, MaxIter):
```

```python
        m,n = p0.shape
        L = np.array([[2,-1,-1],[-1,1,0],[-1,0,1]])
        tL = np.kron(L,np.eye(n))
        Delta_12 = Delta[0]
        Delta_13 = Delta[1]
        q = np.array([Delta_12+Delta_13,-Delta_12,-Delta_13]).reshape(-1,1)
        sigma = 0.3

        p = p0.reshape(-1,1)
        P = p
        u = np.zeros((n*m,1))
        xi = p
        U = u

        for k in range(MaxIter):
            p = proj(p-sigma*(p-xi+u+tL.dot(p)-q), 0, 5)
            u = u+tL.dot(p)-q

            P = np.append(P,p,axis=1)
            U = np.append(U,u,axis=1)
        goal = p.reshape(m,n)
        return goal

class OdomSubscriber(Node):
    def __init__(self):
        super().__init__('odom_subscriber')
        self.declare_parameter('robots_name', '')
        robots_name = self.get_parameter('robots_name').get_parameter_value().
            string_value
        robots_name_list = robots_name.split()

        self.odom_all = [None] * len(robots_name_list)
        for tb3name in robots_name_list:
            self.odom_monitor(tb3name)

    def odom_monitor(self, tb3name):
        robot_id = int(tb3name[-1])
        sub_odom = self.create_subscription(
            Odometry,
            f'/{tb3name}/odom',
            lambda msg: self.odom_call_back(msg, robot_id),
            10)
```

```python
    def odom_call_back(self, msg, robot_id):
        state_x = msg.pose.pose.position.x
        state_y = msg.pose.pose.position.y
        self.odom_all[robot_id-1] = [state_x, state_y]

    def odom_sub_result(self):
        return self.odom_all

class FormationPublisher(Node):
    def __init__(self, goal, robot_number):
        super().__init__('formation_publisher')
        self.robot_number = robot_number
        self.pub_success = False

        self.pub_goal = self.create_publisher(
            Float32MultiArray,
            '/destination',
            1)
        self.pub_goal_data = Float32MultiArray(data=goal)

        self.time_start, _ = self.get_clock().now().seconds_nanoseconds()

        self.timer = self.create_timer(0.01, self.goal_pub)

    def goal_pub(self):
        pub_time = 1  # 最优队形的发布时间设置为1秒
        time_now, _ = self.get_clock().now().seconds_nanoseconds()
        connections = self.pub_goal.get_subscription_count()  # 获取订阅节点的
            个数
        if time_now - self.time_start <= pub_time and connections == self.
            robot_number:
            self.pub_goal.publish(self.pub_goal_data)
        if time_now - self.time_start > pub_time:
            self.pub_success = True

    def goal_pub_result(self):
        return self.pub_success

def main():
    rclpy.init()
    node = rclpy.create_node('formation_ini')
```

```python
        node.declare_parameter('robots_ini', '')
        robots_ini = node.get_parameter('robots_ini').get_parameter_value().
            string_value
        node.destroy_node()

        robot_number = 3
        Delta = np.zeros((3,2,2)) # 存储机器人的相对位置坐标
        Delta[0] = np.array([[1,-1],[1,1]])
        Delta[1] = np.array([[0,-1],[0,1]])
        Delta[2] = np.array([[0,1],[-1,0]])
        Max_Iter = 40 # 算法的最大迭代步数

        goal_str = robots_ini.split()
        goal_float = list(map(float, goal_str))
        goal = np.array(goal_float).reshape(robot_number,-1)

        for i in range(Delta.shape[0]):
            node = OdomSubscriber()
            while rclpy.ok():
                try:
                    rclpy.spin_once(node)
                    odom_all = node.odom_sub_result()
                except KeyboardInterrupt:
                    pass
                if None not in odom_all:
                    distance = np.linalg.norm(goal - np.array(odom_all), axis=1)
                    if (distance < 0.02).all(): # 如果所有机器人都到达目标位置
                        goal = find_optimal(Delta[i], np.array(odom_all), Max_Iter)
                        goal_array = goal.flatten()
                        node.get_logger().info('最优队形: {0}'.format(goal_array))
                        break
            node.destroy_node()

            node = FormationPublisher(goal_array, robot_number)
            while rclpy.ok():
                try:
                    rclpy.spin_once(node)
                    pub_success = node.goal_pub_result()
                except KeyboardInterrupt:
                    pass
                if pub_success:
                    node.get_logger().info("最优队形已经发布，机器人正在移动 ......"
```

```
137                        )
138                    break
139            node.destroy_node()
140        rclpy.shutdown()
141    
142    if __name__ == '__main__':
143        main()
```

下面解释上述代码的主要内容:

- 第 2~6 行加载了运行此文件需要用到的包, 其中第 5 行加载了 ROS 标准消息类型 Float32MultiArray, 可以存储 32 位的浮点数组, 用于存放机器人的最优目标位置;
- 第 8~12 行定义了一个投影函数, 用于限制机器人在空间的移动范围;
- 第 14~36 行实现了集中式算法(2.13), 其中 p_0 是机器人的当前位置坐标, 其值由机器人的里程计数据确定, Delta 是队形矩阵, 函数返回最终计算得到的最优队形位置坐标;
- 第 38~63 行定义了类 OdomSubscriber(), 主要包括以下内容:
  - 定义函数 odom_monitor() 用于启动对所有机器人里程计的订阅, 并将其存储在全局变量 self.odom_all 中;
  - 定义回调函数 odom_call_back() 实时读取里程计数据;
  - 定义函数 odom_sub_result() 用于返回所有机器人的里程计数据.
- 第 65~91 行定义了类 FormationPublisher(), 主要包括以下内容:
  - 第 71~75 行定义了一个发布者用于发布最优队形, 并将发布的数据设置成 Float32MultiArray 类型;
  - 第 79 行定义了一个计时器, 以周期 0.01 秒调用回调函数 goal_pub();
  - 第 81~88 行定义了回调函数 goal_pub(), 用于发布最优队形数据. 这里, 把发布时间设置为 1 秒, 通过 get_subscription_count 方法获取订阅者的数量. 因为采用的是集中式的优化算法, 所以当所有的机器人都连接到发布者时, 发布者才发布最优队形数据, 这样可以保证机器人在队形切换时达到同步;
  - 定义函数 goal_pub_result() 用于返回是否发布成功的逻辑数据.
- 第 93~139 行定义了主函数, 主要包括以下内容:
  - 第 94~98 行对节点进行初始化, 并获取所有机器人的初始位置信息;
  - 第 100~109 行是一些参数设置, 包括机器人数量、队形矩阵、优化算法的最大迭代次数和机器人的初始目标位置, 这里的初始目标位置也即是机器人的初始位置;
  - 第 112~126 行实例化类 OdomSubscriber(), 根据机器人当前的里程计数据计算最优队形, 并将其存储在变量 goal 中;

- 第 128~138 行实例化类 FormationPublisher(), 实现将最优队形数据通过话题 "destination" 发布给所有机器人.

7.2.2 机器人根据最优队形定点移动

接下来, 编写代码实现机器人订阅最优队形, 以及向最优目标位置定点移动. 通过如下命令创建 Python 节点文件:

```
$ cd ~/book_ws/src/chapter7_2/chapter7_2
$ sudo nano tb3_move_to_optimal.py
```

在 tb3_move_to_optimal.py 文件中添加如下代码:

```python
#!/usr/bin/env python3
import rclpy
from rclpy.node import Node
from geometry_msgs.msg import Twist
from nav_msgs.msg import Odometry
from sensor_msgs.msg import LaserScan
from tf_transformations import euler_from_quaternion
from std_msgs.msg import Float32MultiArray
import math
import time

def quat_to_angle(quat):
    rot = euler_from_quaternion((quat.x, quat.y, quat.z, quat.w))
    return rot

def normalize_angle(angle):
    res = angle
    while res > math.pi:
        res -= 2.0*math.pi
    while res < -math.pi:
        res += 2.0*math.pi
    return res

class GetGoal(Node):
    def __init__(self):
        super().__init__('get_goal')
        self.declare_parameter('tb3name', 'turtlebot')
        tb3name = self.get_parameter('tb3name').get_parameter_value().\
            string_value
        self.robot_id = int(tb3name[-1])
```

```python
        self.goal = None
        sub_goal = self.create_subscription(
            Float32MultiArray,
            '/destination',
            self.goal_call_back,
            1)

    def goal_call_back(self, msg):
        if msg is not None:
            self.goal = [msg.data[2*self.robot_id-2], msg.data[2*self.robot_id
                -1]]

    def goal_result(self):
        return self.goal

class RobotControl(Node):
    def __init__(self, goal):
        super().__init__('robot_control')
        self.declare_parameter('tb3name', 'turtlebot')
        tb3name = self.get_parameter('tb3name').get_parameter_value().
            string_value

        self.scan_msg = None
        self.odom_msg = None
        self.is_arrival = False  # 记录机器人是否到达目标位置
        self.goal = goal

        scan_sub = self.create_subscription(
            LaserScan,
            f'/{tb3name}/scan',
            self.scan_call_back,
            1)

        odom_sub = self.create_subscription(
            Odometry,
            f'/{tb3name}/odom',
            self.odom_call_back,
            1)

        self.robot_vel = self.create_publisher(
            Twist,
```

```
                f'/{tb3name}/cmd_vel',
                1)

        self.ini_for_scan()
        self.ini_for_odom()
        robot_timer = self.create_timer(0.01, self.robot_control)

    def ini_for_scan(self):
        while rclpy.ok():
            if self.scan_msg is not None:
                self.get_logger().info('激光雷达完成初始化.')
                break
            rclpy.spin_once(self)
            time.sleep(0.01)

    def ini_for_odom(self):
        while rclpy.ok():
            if self.odom_msg is not None:
                self.get_logger().info('里程计完成初始化.')
                break
            rclpy.spin_once(self)
            time.sleep(0.01)

    def scan_call_back(self, msg):
        self.scan_msg = msg

    def odom_call_back(self, msg):
        self.odom_msg = msg

    def robot_control(self):
        goal_x = self.goal[0]
        goal_y = self.goal[1]
        scan_data = [self.scan_msg.ranges[0], self.scan_msg.ranges[15], self.
            scan_msg.ranges[345]]
        state_x = self.odom_msg.pose.pose.position.x
        state_y = self.odom_msg.pose.pose.position.y
        (roll, pitch, theta) = quat_to_angle(self.odom_msg.pose.pose.
            orientation)

        threshold = 0.4
        tolerance = 0.02
```

```python
            trans = (goal_x - state_x, goal_y - state_y)
            linear = math.sqrt(trans[0] ** 2 + trans[1] ** 2)
            angular = normalize_angle(math.atan2(trans[1], trans[0]) - theta)
            sign = lambda x: math.copysign(1, x) # 定义符号函数
            cmd = Twist()

            if linear > tolerance:
                if scan_data[0] > threshold and scan_data[1] > threshold and
                    scan_data[2] > threshold:
                    if abs(angular) > math.pi/8:
                        cmd.linear.x = min(linear*0.8, 0.22)
                        cmd.angular.z = max(min(angular*0.8, 2.84), -2.84)
                    else:
                        cmd.linear.x = min(linear*1.0, 0.22)
                        cmd.angular.z = max(min(angular*0.1, 2.84), -2.84)
                else:
                    cmd.linear.x = -min(linear*0.1, 0.22)
                    cmd.angular.z = max(min(2.0 * sign(angular), 2.84), -2.84)
                    if scan_data[0] > threshold and scan_data[1] > threshold and
                        scan_data[2] > threshold:
                        cmd.linear.x = 0.1
                        cmd.angular.z = 0.0
            else:
                cmd.linear.x = 0.0
                cmd.angular.z = 0.0
                self.is_arrival = True
            self.robot_vel.publish(cmd)

    def arrival_result(self):
        return self.is_arrival

def main():
    rclpy.init()

    while rclpy.ok():
        node = GetGoal()
        while rclpy.ok():
            try:
                rclpy.spin_once(node)
                goal = node.goal_result()
            except KeyboardInterrupt:
                pass
```

```
150            if goal is not None:
151                node.get_logger().info('目标位置: {0}'.format(goal))
152                break
153        node.destroy_node()
154
155        node = RobotControl(goal)
156        while rclpy.ok():
157            try:
158                rclpy.spin_once(node)
159                is_arrival = node.arrival_result()
160            except KeyboardInterrupt:
161                pass
162            if is_arrival:
163                break
164        node.destroy_node()
165    rclpy.shutdown()
166
167 if __name__ == '__main__':
168     main()
```

下面解释上述代码的主要内容:

- 第 2~10 行加载了运行此文件需要用到的包, 其中第 6 行加载了 ROS 传感器消息类型 LaserScan, 用于控制机器人移动时进行避障处理;
- 第 24~43 行定义了类 GetGoal(), 主要包括以下内容:
  - 第 31~36 行订阅中心节点发布的最优队形话题 "destination", 用于确定机器人的目标位置;
  - 第 38~40 行定义的回调函数 goal_call_back() 对最优队形话题的数据进行处理. 因为中心节点将所有机器人的最优目标位置发布给了每一个机器人, 因此机器人需要根据自己的名称编号对最优目标位置进行切片处理.
- 第 45~137 行定义了类 RobotControl(), 主要包括以下内容:
  - 第 56~66 行对机器人的激光雷达和里程计订阅进行初始化, 并在第 73~74 行完成对激光雷达和里程计的初始化;
  - 第 99~134 行定义了对机器人进行速度控制的函数 robot_control(). 这个函数和 7.1.1 节的代码相似, 不同的地方是在这里增加了激光雷达数据, 并对机器人做简单的避障处理. 通过读取机器人正前方左右各 15° 的范围, 判断机器人前方是否有障碍物. 如果有障碍物, 就减小线速度, 同时增加角速度.
- 第 139~165 行定义了主函数, 主要是对类 GetGoal() 和类 RobotControl() 先后交替进行实例化, 即在确定了机器人的目标位置后, 对机器人进行速度控制.

打开 chapter7_2 目录下的 setup.py 文件, 将 entry_points 域修改为如下形式:

```
1  entry_points={
2      'console_scripts': [
3          'optimal_formation_publisher = chapter7_2.optimal_formation_publisher:
              main',
4          'tb3_move_to_optimal = chapter7_2.tb3_move_to_optimal:main',
5      ],
6  },
```

构建 chapter7_2 功能包，运行如下命令：

```
$ cd ~/book_ws
$ colcon build --packages-select chapter7_2
```

7.2.3 编写 launch 文件测试队形切换效果

在这一节中，编写一个 launch 文件测试上述队形切换代码的效果；在 chapter7_2 目录下新建一个 launch 目录，用于存放 launch 文件，运行如下命令：

```
$ cd ~/book_ws/src/chapter7_2
$ mkdir launch
```

在 launch 目录下新建一个名称为 tb3_formation_demo.launch.py 的文件：

```
$ cd ~/book_ws/src/chapter7_2/launch
$ sudo nano tb3_formation_demo.launch.py
```

并添加如下代码：

```
1   #!/usr/bin/env python3
2   import os
3   from ament_index_python.packages import get_package_share_directory
4   from launch import LaunchDescription
5   from launch.actions import IncludeLaunchDescription
6   from launch.launch_description_sources import PythonLaunchDescriptionSource
7   from launch_ros.actions import Node
8
9   def generate_launch_description():
10      TURTLEBOT3_MODEL = os.environ['TURTLEBOT3_MODEL']
11      model_folder = 'turtlebot3_' + TURTLEBOT3_MODEL
12      sdf_path = os.path.join(
13          get_package_share_directory('turtlebot3_gazebo'),
14          'models',
15          model_folder,
```

```python
                'model.sdf'
            )

    urdf_file_name = 'turtlebot3_' + TURTLEBOT3_MODEL + '.urdf'
    urdf_path = os.path.join(
        get_package_share_directory('turtlebot3_description'),
        'urdf',
        urdf_file_name)
    with open(urdf_path, 'r') as infp:
        robot_description = infp.read()

    default_world_path = os.path.join(
        get_package_share_directory('turtlebot3_gazebo'),
        'worlds',
        'empty_world.world'
        )

    robots = [
        {'name': 'robot1', 'x_pose': '0.0', 'y_pose': '0.0', 'z_pose': '0.01'
            },
        {'name': 'robot2', 'x_pose': '0.0', 'y_pose': '1.0', 'z_pose': '0.01'
            },
        {'name': 'robot3', 'x_pose': '0.0', 'y_pose': '2.0', 'z_pose': '0.01'}
        ]

    gazebo_load_cmd = IncludeLaunchDescription(
        PythonLaunchDescriptionSource(
            os.path.join(get_package_share_directory('gazebo_ros'), 'launch',
                'gazebo.launch.py')
            ),
        launch_arguments={'world': default_world_path}.items()
        )

    ld = LaunchDescription()
    ld.add_action(gazebo_load_cmd)

    robots_name = ''
    robots_ini = ''
    for i in range(len(robots)):
        spawn_robot_cmd = Node(
            package = 'gazebo_ros',
            executable = 'spawn_entity.py',
```

```python
            name = robots[i].get('name')+'_spawn_node',
            output = 'screen',
            arguments = [
                '-entity', robots[i].get('name')+'_entity',
                '-robot_namespace', robots[i].get('name'),
                '-file', sdf_path,
                '-x', robots[i].get('x_pose'),
                '-y', robots[i].get('y_pose'),
                '-z', robots[i].get('z_pose'),
            ],
        )
        ld.add_action(spawn_robot_cmd)

        robot_state_publisher_cmd = Node(
            package = 'robot_state_publisher',
            executable = 'robot_state_publisher',
            namespace = robots[i].get('name'),
            name = robots[i].get('name')+'_state_node',
            output = 'screen',
            parameters = [{
                'use_sim_time': True,
                'robot_description': robot_description,
                'publish_frequency': 50.0
            }],
        )
        ld.add_action(robot_state_publisher_cmd)

        multi_tb3_move_to_goal_cmd = Node(
            package = 'chapter7_2',
            executable = 'tb3_move_to_optimal',
            name = robots[i].get('name')+'_move_to_optimal_node',
            output = 'screen',
            parameters = [{
                'tb3name': robots[i].get('name')
            }],
        )
        ld.add_action(multi_tb3_move_to_goal_cmd)

        robots_name = robots_name + robots[i].get('name') + ' '
        robots_ini = robots_ini + robots[i].get('x_pose') + ' ' + robots[i].get('y_pose') + ' '

```

```
 96     optimal_formation_publisher_cmd = Node(
 97         package = 'chapter7_2',
 98         executable = 'optimal_formation_publisher',
 99         name = 'optimal_formation_publisher_node',
100         output = 'screen',
101         parameters = [{
102             'robots_name': robots_name,
103             'robots_ini': robots_ini
104             }],
105         )
106     ld.add_action(optimal_formation_publisher_cmd)
107
108     return ld
```

这里的代码与第7.1.2节相似, 不同的地方解释如下:

- 第 49~50 行定义了两个变量分别用于存储所有机器人的名称和初始位置坐标, 并且在代码的第 93~94 行进行赋值;
- 第 82~91 行加载前面编写的机器人控制程序, 实现对机器人从当前位置移动到最优目标位置的速度控制;
- 第 96~106 行加载前面编写的最优队形发布程序, 实现集中式算法求解最优队形并将其发布给所有机器人.

重新打开 chapter7_2 目录下的 setup.py 文件, 将 data_files 域修改为如下形式:

```
1  data_files=[
2      ...
3      (os.path.join('share', package_name, 'launch'), glob(os.path.join('launch
       ', '*.launch.py'))),
4  ],
```

并在文件头部添加如下代码:

```
1  import os
2  from glob import glob
```

接下来, 重新构建 chapter7_2 功能包, 运行如下命令:

```
$ cd ~/book_ws
$ colcon build --packages-select chapter7_2
```

打开终端, 输入如下命令进行测试:

```
$ ros2 launch chapter7_2 tb3_formation_demo.launch.py
```

图7.5给出了在 Gazebo 中的仿真结果, 可以看到机器人从直线队形出发, 首先切换到三角形队形, 再切换到直线队形, 最后又切换到三角形队形.

注 这里没有考虑机器人与最优队形的分配问题, 即借助组合优化方法将机器人的顺序进行调整, 可以减小总的移动距离. 感兴趣的读者可以阅读参考文献 [8].

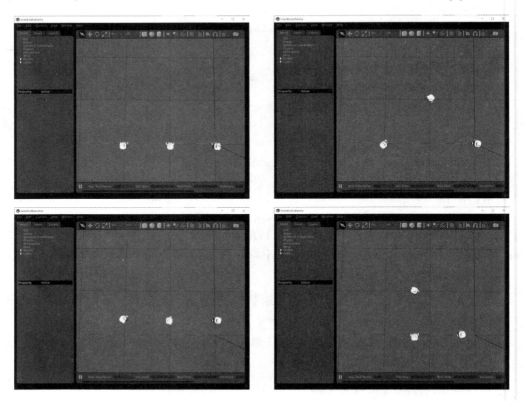

图 7.5　在 Gazebo 仿真中三个 TurtleBot 3 机器人的队形切换

7.3　多个机器人队形切换——分布式优化算法

在上一节的程序设计中, 有一个中心节点用于订阅所有机器人的里程计话题并计算机器人的最优队形. 从图7.3可以看到, 机器人只与中心节点进行通信, 机器人之间没有任何通信. 这一节中将删除中心节点, 实现第2章中公式(2.21)所表示的分布式优化算法.

同样考虑图7.2中所示的三个队形的切换问题. 设计框架如下: 三个机器人的通信拓扑关系和机器人间的相对位置关系保持一致, 即 robot1 与 robot2 和 robot3 通信, robot2 只与 robot1 通信, robot3 也只与 robot1 通信. 此设计框架如图7.6所示, 用一句话概括, 即是分布式计算分布式控制.

同样地, 将编写代码实现上述队形切换任务, 算法流程如图7.7所示. 可以看出该流程图主要包括一个程序文件 tb3_distributed_formation.py, 每个机器人订阅邻居的里程计话题, 分布式计算最优队形, 同时控制自己移动到最优目标位置. 接下来, 将具体编写这个程序文件.

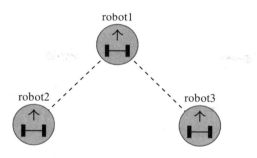

图 7.6　分布式计算分布式控制框架

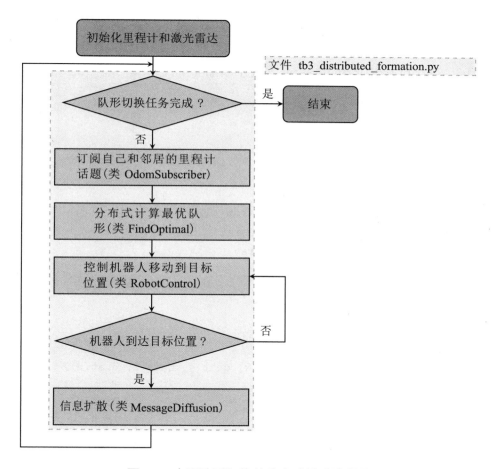

图 7.7　实现队形切换的分布式算法流程图

执行如下命令创建一个名称为 chapter7_3 的功能包：

```
$ cd ~/book_ws/src
$ ros2 pkg create --build-type ament_python chapter7_3
```

然后构建这个包：

```
$ cd ~/book_ws
$ colcon build --packages-select chapter7_3
$ . install/setup.bash
```

7.3.1 分布式计算最优队形和机器人控制

在 chapter7_3 功能包中创建一个同时实现分布式优化计算和控制机器人定点移动的 Python 节点文件:

```
$ cd ~/book_ws/src/chapter7_3/chapter7_3
$ sudo nano tb3_distributed_formation.py
```

在 tb3_distributed_formation.py 文件中添加如下代码:

```python
#!/usr/bin/env python3
import rclpy
from rclpy.node import Node
from geometry_msgs.msg import Twist
from nav_msgs.msg import Odometry
from sensor_msgs.msg import LaserScan
from tf_transformations import euler_from_quaternion
from std_msgs.msg import Float32MultiArray
from std_msgs.msg import Int32MultiArray
import numpy as np
import math
import time

def quat_to_angle(quat):
    rot = euler_from_quaternion((quat.x, quat.y, quat.z, quat.w))
    return rot

def normalize_angle(angle):
    res = angle
    while res > math.pi:
        res -= 2.0*math.pi
    while res < -math.pi:
        res += 2.0*math.pi
    return res

def proj(x, lb, ub):
    y = x.copy()
    y[x>ub] = ub
```

```python
29            y[x<lb] = lb
30            return y
31
32    class OdomSubscriber(Node):
33        def __init__(self):
34            super().__init__('odom_subscriber')
35            self.declare_parameter('tb3name', '')
36            self.declare_parameter('neighbor_name', '')
37            self.tb3name = self.get_parameter('tb3name').get_parameter_value().
                  string_value
38            neighbor_name = self.get_parameter('neighbor_name').
                  get_parameter_value().string_value
39
40            self_neighbor_list = neighbor_name.split()
41            self_neighbor_list.insert(0, self.tb3name) # 将机器人自身的名字加在首位
42
43            self.odom_all = [None] * len(self_neighbor_list)
44            for index, name in enumerate(self_neighbor_list):
45                self.odom_monitor(index, name) # 订阅自身和邻居的里程计
46
47        def odom_monitor(self, index, name):
48            sub_odom = self.create_subscription(
49                Odometry,
50                f'/{name}/odom',
51                lambda msg: self.odom_call_back(msg, index),
52                10)
53
54        def odom_call_back(self, msg, index):
55            state_x = msg.pose.pose.position.x
56            state_y = msg.pose.pose.position.y
57            self.odom_all[index] = [state_x, state_y]
58
59        def odom_sub_result(self):
60            return self.odom_all
61
62    class FindOptimal(Node):
63        def __init__(self, odom, Delta):
64            super().__init__('find_optimal')
65            self.declare_parameter('tb3name', '')
66            self.declare_parameter('neighbor_name', '')
67            self.tb3name = self.get_parameter('tb3name').get_parameter_value().
                  string_value
```

```
68          neighbor_name = self.get_parameter('neighbor_name').
                get_parameter_value().string_value
69          neighbor_list = neighbor_name.split()
70          self.robot_id = int(self.tb3name[-1]) # 取机器人名称的最后一个字符作为
                robot_id，即取值为1、2、3
71
72          Delta_12 = Delta[0]
73          Delta_13 = Delta[1]
74          self.q_all = np.array([Delta_12+Delta_13,-Delta_12,-Delta_13])
75
76          self.xi = np.array(odom[0]).reshape(-1,1) # 初始位置常量\xi_i
77          self.q_i = self.q_all[self.robot_id-1].reshape(-1,1) # 分布式算法中的
                常量q_i
78          self.current_opt = np.array(odom[0]).reshape(-1,1) # 优化变量p_i初始值
79          self.u_i = np.zeros((2,1)) # 辅助变量u_i初始值
80
81          self.p_pub = self.current_opt.flatten()
82          self.pub_opt = self.create_publisher( # 定义发布者用于发布优化变量p_i
83              Float32MultiArray,
84              f'/{self.tb3name}/opt_find',
85              1)
86
87          self.opt_neighbor = odom[1:] # 用邻居的里程计数据初始化邻居的最优解
88          for index, name in enumerate(neighbor_list):
89              self.opt_monitor(index, name) # 初始化函数opt_monitor
90
91          self.timer = self.create_timer(0.1, self.opt_pub) # 优化算法求解并发布
92
93      def opt_monitor(self, index, name): # 订阅邻居的优化变量p_j
94          sub_opt = self.create_subscription(
95              Float32MultiArray,
96              f'/{name}/opt_find',
97              lambda msg: self.opt_call_back(msg, index),
98              1)
99
100     def opt_call_back(self, msg, index):
101         self.opt_neighbor[index] = msg.data
102
103     def opt_pub(self):
104         sigma = 0.3
105         xi_i = self.xi
106         q_i = self.q_i
```

```python
            p_i = self.current_opt
            u_i = self.u_i

            opt_neighbor = np.array(self.opt_neighbor)
            p_ij = np.zeros((2,1))
            for index in range(len(opt_neighbor)):
                p_ij += p_i-opt_neighbor[index].reshape(-1,1) # 自身与邻居状态差的
                    累加

            p_i = proj(p_i-sigma*(p_i-xi_i+u_i+p_ij-q_i), 0, 5) # 分布式算法的迭代
                公式
            u_i = u_i+p_ij-q_i

            self.current_opt = p_i
            self.u_i = u_i
            self.p_pub = self.current_opt.flatten()
            self.pub_opt_data = Float32MultiArray(data=self.p_pub)

            connections = self.pub_opt.get_subscription_count()
            if connections == len(opt_neighbor): # 如果所有的邻居连接成功，则发布当
                前的优化变量值
                self.pub_opt.publish(self.pub_opt_data)

    def opt_result(self):
        return self.tb3name, self.current_opt.flatten()

class RobotControl(Node):
    def __init__(self, goal):
        super().__init__('robot_control')
        self.declare_parameter('tb3name', 'turtlebot')
        tb3name = self.get_parameter('tb3name').get_parameter_value().
            string_value

        self.scan_msg = None
        self.odom_msg = None
        self.is_arrival = False # 记录机器人是否到达目标位置
        self.goal = goal

        scan_sub = self.create_subscription(
            LaserScan,
            f'/{tb3name}/scan',
            self.scan_call_back,
```

```
145                1)
146
147        odom_sub = self.create_subscription(
148            Odometry,
149            f'/{tb3name}/odom',
150            self.odom_call_back,
151            1)
152
153        self.robot_vel = self.create_publisher(
154            Twist,
155            f'/{tb3name}/cmd_vel',
156            1)
157
158        self.ini_for_scan()
159        self.ini_for_odom()
160        robot_timer = self.create_timer(0.01, self.robot_control)
161
162    def ini_for_scan(self):
163        while rclpy.ok():
164            if self.scan_msg is not None:
165                self.get_logger().info('激光雷达完成初始化.')
166                break
167            rclpy.spin_once(self)
168            time.sleep(0.01)
169
170    def ini_for_odom(self):
171        while rclpy.ok():
172            if self.odom_msg is not None:
173                self.get_logger().info('里程计完成初始化.')
174                break
175            rclpy.spin_once(self)
176            time.sleep(0.01)
177
178    def scan_call_back(self, msg):
179        self.scan_msg = msg
180
181    def odom_call_back(self, msg):
182        self.odom_msg = msg
183
184    def robot_control(self):
185        goal_x = self.goal[0]
186        goal_y = self.goal[1]
```

```python
        scan_data = [self.scan_msg.ranges[0], self.scan_msg.ranges[15], self.
            scan_msg.ranges[345]]
        state_x = self.odom_msg.pose.pose.position.x
        state_y = self.odom_msg.pose.pose.position.y
        (roll, pitch, theta) = quat_to_angle(self.odom_msg.pose.pose.
            orientation)

        threshold = 0.4 # 激光雷达防碰撞的敏感值
        tolerance = 0.02 # 到达目标位置允许的误差值

        trans = (goal_x - state_x, goal_y - state_y)
        linear = math.sqrt(trans[0] ** 2 + trans[1] ** 2)
        angular = normalize_angle(math.atan2(trans[1], trans[0]) - theta)
        sign = lambda x: math.copysign(1, x) # 定义符号函数
        cmd = Twist()

        if linear > tolerance: # 机器人移动和避障控制
            if scan_data[0] > threshold and scan_data[1] > threshold and
                scan_data[2] > threshold:
                if abs(angular) > math.pi/8:
                    cmd.linear.x = min(linear*0.8, 0.22)
                    cmd.angular.z = max(min(angular*0.8, 2.84), -2.84)
                else:
                    cmd.linear.x = min(linear*1.0, 0.22)
                    cmd.angular.z = max(min(angular*0.1, 2.84), -2.84)
            else:
                cmd.linear.x = -min(linear*0.1, 0.22)
                cmd.angular.z = max(min(2.0 * sign(angular), 2.84), -2.84)
                if scan_data[0] > threshold and scan_data[1] > threshold and
                    scan_data[2] > threshold:
                    cmd.linear.x = 0.1
                    cmd.angular.z = 0.0
        else:
            cmd.linear.x = 0.0
            cmd.angular.z = 0.0
            self.is_arrival = True
        self.robot_vel.publish(cmd)

    def arrival_result(self):
        return self.is_arrival


class MessageDiffusion(Node):
```

```python
225      def __init__(self):
226          super().__init__('message_diffusion')
227          self.declare_parameter('tb3name', '')
228          self.declare_parameter('neighbor_name', '')
229          tb3name = self.get_parameter('tb3name').get_parameter_value().
                 string_value
230          neighbor_name = self.get_parameter('neighbor_name').
                 get_parameter_value().string_value
231          self.neighbor_list = neighbor_name.split()
232          robot_id = int(tb3name[-1])
233
234          self.msg_pub = [robot_id] # 记录发布的信息
235          self.pub_arriver = self.create_publisher( # 定义发布者用于发布到达目标
                 位置的确认信息
236              Int32MultiArray,
237              f'/{tb3name}/arriver_confirm',
238              10)
239
240          for name in self.neighbor_list:
241              self.arriver_monitor(name)
242
243          arriver_pub_timer = self.create_timer(0.1, self.arriver_pub) # 到达目
                 标位置后发布确认信息
244
245      def arriver_monitor(self, name): # 订阅邻居的确认信息
246          sub_arriver = self.create_subscription(
247              Int32MultiArray,
248              f'/{name}/arriver_confirm',
249              self.arriver_call_back,
250              1)
251
252      def arriver_call_back(self, msg):
253          self.msg_pub = list(set(self.msg_pub+list(msg.data)))
254
255      def arriver_pub(self):
256          self.pub_arriver_data = Int32MultiArray(data=self.msg_pub)
257          connections = self.pub_arriver.get_subscription_count()
258          if connections == len(self.neighbor_list):
259              self.pub_arriver.publish(self.pub_arriver_data)
260
261      def arriver_result(self, robot_number):
262          all_arrival = False
```

```python
            if len(self.msg_pub) == robot_number:  # 如果所有机器人都到达目标位置，
                                                   # 则返回逻辑值True
                all_arrival = True
        return all_arrival

def main():
    rclpy.init()
    robot_number = 3
    Delta = np.zeros((3,2,2))
    Delta[0] = np.array([[1,-1],[1,1]])  # 队形约束是robot1与robot2和robot3的相
                                          # 对位置
    Delta[1] = np.array([[0,-1],[0,1]])
    Delta[2] = np.array([[0,1],[-1,0]])

    for i in range(Delta.shape[0]):
        node = OdomSubscriber()
        while rclpy.ok():
            try:
                rclpy.spin_once(node)
                odom = node.odom_sub_result()
            except KeyboardInterrupt:
                pass
            if None not in odom:
                break
        node.destroy_node()

        node = FindOptimal(odom, Delta[i])
        time_start, _ = node.get_clock().now().seconds_nanoseconds()
        while rclpy.ok():
            try:
                rclpy.spin_once(node)
                tb3name, goal = node.opt_result()
            except KeyboardInterrupt:
                pass
            time_now, _ = node.get_clock().now().seconds_nanoseconds()
            if time_now - time_start > 3:
                node.get_logger().info('{0}得到的最优解：{1}'.format(tb3name,
                    goal))
                break
        node.destroy_node()

        node = RobotControl(goal)
```

```
302          while rclpy.ok():
303              try:
304                  rclpy.spin_once(node)
305                  is_arrival = node.arrival_result()
306              except KeyboardInterrupt:
307                  pass
308              if is_arrival:
309                  break
310          node.destroy_node()
311
312          node = MessageDiffusion()
313          delay_counter = 0 # 延迟计数器
314          while rclpy.ok():
315              try:
316                  rclpy.spin_once(node)
317                  all_arrival = node.arriver_result(robot_number)
318              except KeyboardInterrupt:
319                  pass
320              if all_arrival is True:
321                  delay_counter += 1
322                  if delay_counter == 20:
323                      node.get_logger().info('所有机器人都已到达目标位置.')
324                      break
325          node.destroy_node()
326      rclpy.shutdown()
327
328  if __name__ == '__main__':
329      main()
```

下面解释上述代码的主要内容:

- 第 2~12 行加载了运行此文件需要用到的包, 其中第 9 行加载了 ROS 标准消息类型 Int32MultiArray, 将在类 MessageDiffusion() 中使用, 用于机器人到达目标位置后进行信息扩散, 以确认所有的机器人都到达最优目标位置;
- 第 32~60 行定义了类 OdomSubscriber(), 其内容和第 7.2.1 节定义的同名类的内容相似, 区别是这里的每个机器人订阅的都是自己和邻居的里程计话题;
- 第 62~128 行定义了类 FindOptimal(), 主要包括以下内容:
  - 第 76~79 行分别对算法公式(2.21)中的常量 ξ 和 q_i 赋值, 对变量 p_i 和 u_i 赋初值. 机器人间的连接权重 a_{ij}, 当两个机器人有通信时取值 1, 否则取值 0;
  - 第 81~85 行定义了一个发布者, 用于发布优化变量 p_i 的值, 实现分布式计算;

- 第 91 行定义了一个计时器, 以周期 0.1 秒调用回调函数 opt_pub(), 计算最优解并发布;
- 第 93~98 行定义了函数 opt_monitor(), 用于订阅邻居在优化计算过程中发布的变量 p_j 的值;
- 第 103~125 行实现分布式优化算法计算最优队形, 其中第 110~113 行获取邻居的变量 p_j 的值, 并计算公式(2.21)中的求和部分 $\sum_{j \in N_i} a_{ij}(p_i(k) - p_j(k))$, 第 115~116 行实现算法公式(2.21)中的两个迭代式, 第 118~125 行将优化算法计算过程中得到的变量 p_i 的值发布给自己的邻居.
- 第 130~222 行定义了类 RobotControl(), 与第7.2.2节的同名类代码相同;
- 第 224~265 行定义了类 MessageDiffusion(), 用于确认是否所有的机器人都到达了目标位置, 主要包括以下内容:
  - 第 234~238 行初始化一个发布者, 用于发布已经到达目标位置的机器人信息, 即机器人自身的编号 robot_id;
  - 第 240~241 行初始化函数 arriver_monitor(), 其内容在第 245~250 行给出, 用于订阅机器人邻居到达目标位置时所发布的确认信息;
  - 第 252~253 行定义了回调函数 arriver_call_back(), 将机器人自己发布的确认信息和接收到的邻居的确认信息整合到一个列表中, 即获得当前网络中已经到达目标位置的机器人编号, 并将该信息在第 255~259 行定义的函数 arriver_pub() 中发布给邻居;
  - 第 261~265 行定义了函数 arriver_result(), 当机器人自己能够确认网络中的所有机器人都到达目标位置时, 返回逻辑值 True.
- 第 267~326 行定义了主函数, 主要包括以下内容:
  - 第 268~273 行对节点进行初始化和对参数进行设置, 包括机器人数量和队形矩阵;
  - 第 275~285 行实例化类 OdomSubscriber(), 主要用于订阅机器人自己和邻居的里程计话题;
  - 第 287~299 行实例化类 FindOptimal(), 用于计算最优队形, 并将机器人自己的最优目标位置存储在变量 goal 中. 这里我们设置的优化算法求解时间是 3 秒;
  - 第 301~310 行实例化类 RobotControl(), 用于对机器人进行速度控制;
  - 第 312~325 行实例化类 MessageDiffusion(), 用于确定是否网络中的所有机器人都已到达目标位置. 这里使用了一个延时计数器, 用于处理网络通信延时产生的错误.

打开 chapter7_3 目录下的 setup.py 文件, 将 entry_points 域修改为如下形式:

```python
entry_points={
    'console_scripts': [
        'tb3_distributed_formation = chapter7_3.tb3_distributed_formation:main
        ',
    ],
},
```

7.3.2 编写 launch 文件测试队形切换效果

在这一节中，编写一个 launch 文件测试上述队形切换代码的效果；在 chapter7_3 目录下新建一个 launch 目录，用于存放 launch 文件，运行如下命令：

```
$ cd ~/book_ws/src/chapter7_3
$ mkdir launch
```

在 launch 目录下新建一个名称为 tb3_distributed_formation_demo.launch.py 的文件：

```
$ cd ~/book_ws/src/chapter7_3/launch
$ sudo nano tb3_distributed_formation_demo.launch.py
```

并添加如下代码：

```python
#!/usr/bin/env python3
import os
from ament_index_python.packages import get_package_share_directory
from launch import LaunchDescription
from launch.actions import IncludeLaunchDescription
from launch.launch_description_sources import PythonLaunchDescriptionSource
from launch_ros.actions import Node

def generate_launch_description():
    TURTLEBOT3_MODEL = os.environ['TURTLEBOT3_MODEL']
    model_folder = 'turtlebot3_' + TURTLEBOT3_MODEL
    sdf_path = os.path.join(
        get_package_share_directory('turtlebot3_gazebo'),
        'models',
        model_folder,
        'model.sdf'
        )
    
    urdf_file_name = 'turtlebot3_' + TURTLEBOT3_MODEL + '.urdf'
```

```python
urdf_path = os.path.join(
    get_package_share_directory('turtlebot3_description'),
    'urdf',
    urdf_file_name)
with open(urdf_path, 'r') as infp:
    robot_description = infp.read()

default_world_path = os.path.join(
    get_package_share_directory('turtlebot3_gazebo'),
    'worlds',
    'empty_world.world'
    )

robots = [
    {'name': 'robot1', 'x_pose': '0.0', 'y_pose': '0.0', 'z_pose': '0.01',
        'neighbor': 'robot2 robot3'},
    {'name': 'robot2', 'x_pose': '0.0', 'y_pose': '1.0', 'z_pose': '0.01',
        'neighbor': 'robot1'},
    {'name': 'robot3', 'x_pose': '0.0', 'y_pose': '2.0', 'z_pose': '0.01',
        'neighbor': 'robot1'}
    ]

gazebo_load_cmd = IncludeLaunchDescription(
    PythonLaunchDescriptionSource(
        os.path.join(get_package_share_directory('gazebo_ros'), 'launch',
            'gazebo.launch.py')
        ),
    launch_arguments = {'world': default_world_path}.items()
    )

ld = LaunchDescription()
ld.add_action(gazebo_load_cmd)

for i in range(len(robots)):
    spawn_robot_cmd = Node(
        package = 'gazebo_ros',
        executable = 'spawn_entity.py',
        name = robots[i].get('name')+'_spawn_node',
        output = 'screen',
        arguments = [
            '-entity', robots[i].get('name')+'_entity',
            '-robot_namespace', robots[i].get('name'),
```

```
58              '-file', sdf_path,
59              '-x', robots[i].get('x_pose'),
60              '-y', robots[i].get('y_pose'),
61              '-z', robots[i].get('z_pose'),
62          ],
63      )
64      ld.add_action(spawn_robot_cmd)
65
66      robot_state_publisher_cmd = Node(
67          package = 'robot_state_publisher',
68          executable = 'robot_state_publisher',
69          namespace = robots[i].get('name'),
70          name = robots[i].get('name')+'_state_node',
71          output = 'screen',
72          parameters = [{
73              'use_sim_time': True,
74              'robot_description': robot_description,
75              'publish_frequency': 50.0
76          }],
77      )
78      ld.add_action(robot_state_publisher_cmd)
79
80      tb3_distributed_formation_cmd = Node(
81          package = 'chapter7_3',
82          executable = 'tb3_distributed_formation',
83          name = robots[i].get('name')+'_node',
84          output = 'screen',
85          parameters = [{
86              'tb3name': robots[i].get('name'),
87              'neighbor_name': robots[i].get('neighbor')
88          }],
89      )
90      ld.add_action(tb3_distributed_formation_cmd)
91
92  return ld
```

这里的代码与7.2.3相似, 对不同的地方解释如下:

- 第 33~37 行定义了变量 robots 用于存储所有机器人的信息, 包括机器人的名称、初始位置和邻居的名称;
- 第 80~90 行加载前面编写的分布式编队的程序, 实现对优化问题的分布式计算和对机器人的分布式控制.

重新打开 chapter7_3 目录下的 setup.py 文件，将 data_files 域修改为如下形式：

```
1  data_files=[
2      ...
3      (os.path.join('share', package_name, 'launch'), glob(os.path.join('launch
       ', '*.launch.py'))),
4  ],
```

并在文件头部添加如下代码：

```
1  import os
2  from glob import glob
```

接下来，重新构建 chapter7_3 功能包，运行如下命令：

```
$ cd ~/book_ws
$ colcon build --packages-select chapter7_3
```

打开终端，输入如下命令进行测试：

```
$ ros2 launch chapter7_3 tb3_distributed_formation_demo.launch.py
```

图7.8给出的在 Gazebo 中的仿真结果和图7.5相似. 不同的是这里采用了分布式算法求解优化问题，由于机器人之间通信的延时产生了优化求解的不同步性，最后得到的最优解和集中式算法得到的最优解可能会有少许区别，在仿真结果上表现为得到的队形空间位置可能会不同.

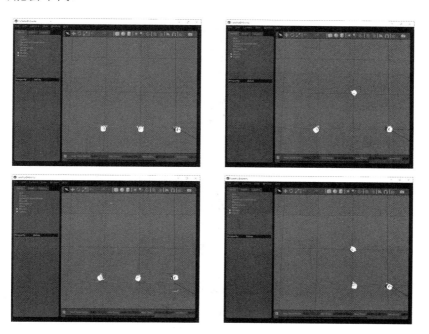

图 7.8 在 **Gazebo** 仿真中三个 **TurtleBot 3** 机器人的队形切换

可以通过如下命令可视化节点和话题的关系图:

```
$ rqt_graph
```

图 7.9 是截取的三个机器人进行分布式优化计算时的节点和话题关系图. 从图中可以看到三个机器人进行分布式优化计算时对话题 "opt_find" 的发布和订阅关系: 节点 "/robot1_node" 发布话题 "/robot1/opt_find", 同时订阅话题 "/robot2/opt_find" 和 "/robot3/opt_find", 这是由于 robot1 有两个邻居 robot2 和 robot3; 节点 "/robot2_node" 发布话题 "/robot2/opt_find", 同时订阅话题 "/robot1/opt_find", 这是由于 robot2 只有一个邻居 robot1; 节点 "/robot3_node" 发布话题 "/robot3/opt_find", 同时订阅话题 "/robot1/opt_find", 这是由于 robot3 也只有一个邻居 robot1. 上述的通信关系明确反映了所设计的分布式计算框架, 即和图 7.6 的通信拓扑结构保持一致.

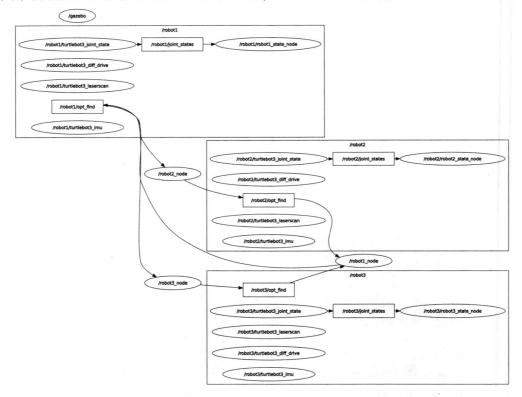

图 7.9 三个机器人进行分布式优化计算时的节点和话题关系图

练 习

1. 参考第 7.1 节的代码, 练习实现机器人以原点为中心、半径为 2 的圆周运动.
2. 参考第 7.2 节和 7.3 节的代码, 分别实现五个机器人的队形切换任务.

第 8 章　基于几何相似性的多机器人编队仿真

第3章介绍了基于几何相似性的多机器人编队问题. 基于几何相似性的队形具有更大的灵活性, 多个机器人形成的队形可以在旋转、伸缩和平移变换下保持相似性. 本章将基于第3章所述的利用几何相似性所设计的多机器人编队算法, 在 ROS 2 的 Gazebo 仿真中实现多机器人的队形切换.

8.1　算法设计框架

在这一章中, 将在 Gazebo 仿真中实现四个 Turtlebot 3 机器人的队形切换任务. 四个机器人组成的三个队形图标如图8.1所示. 四个机器人从直线队形出发, 依次实现图8.1中从左往右的三个队形的切换.

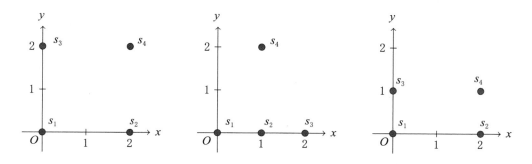

图 8.1　本章中考虑的四个机器人形成的三个队形图标

为了实现上述的队形切换, 考虑第3章中的分布式优化算法. 四个机器人形成一个连通的通信拓扑结构, 如图8.2所示, 其中 robot1~robot4 分别与图8.1中的 s_1~s_4 对应. 图中的虚线表示机器人之间的通信关系. 从图中我们可以看出, robot1 的邻居是 robot3, robot2 的邻居是 robot3 和 robot4, robot3 的邻居是 robot1 和 robot2, robot4 的邻居是 robot2.

将编写代码实现上述队形切换任务, 算法流程如图8.3所示. 该流程图主要包括三个程序文件, 其中 robot1_distributed_formation.py 用于实现 robot1 的分布式计算和控制, robot2_distributed_formation.py 用于实现 robot2 的分布式计算和控制,

roboti_distributed_formation.py 用于实现 robot3 和 robot4 的分布式计算和控制. 类 OdomSubscriber() 实现订阅机器人自己、robot1 和 robot2 的里程计数据,用于初始化分布式优化算法(3.25)~(3.27)的变量 p_1, p_2 和 p_i,即从机器人当前的位置开始计算下一个最优队形,避免使用随机初始化.

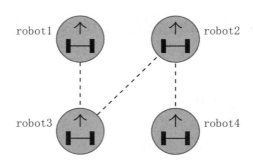

图 8.2　四个机器人形成连通的通信拓扑结构

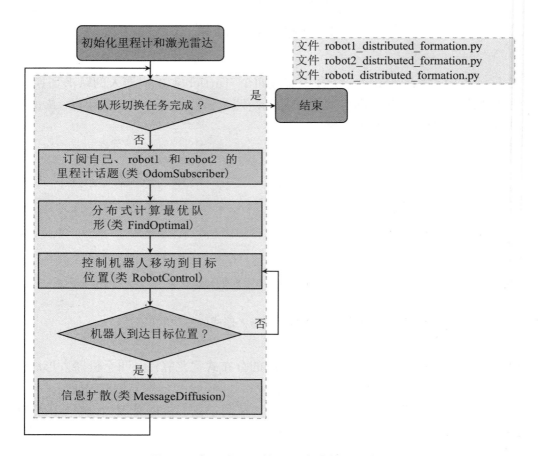

图 8.3　实现队形切换的分布式算法流程图

8.2 代码实现

在这一节中,将具体编写实现上述设计框架的程序文件,执行如下命令创建一个名称为 chapter8_2 的功能包:

```
$ cd ~/book_ws/src
$ ros2 pkg create --build-type ament_python chapter8_2
```

然后构建这个包:

```
$ cd ~/book_ws
$ colcon build --packages-select chapter8_2
$ . install/setup.bash
```

8.2.1 节点代码实现

接下来,将基于第3章的公式(3.25)~(3.27)分别编写相应的分布式优化算法. 由于这三个算法迭代公式有少许差别,因此需要编写三个不同的节点文件.

在 chapter8_2 功能包中创建实现 robot1 分布式优化计算和移动控制的 Python 节点文件:

```
$ cd ~/book_ws/src/chapter8_2/chapter8_2
$ sudo nano robot1_distributed_formation.py
```

在 robot1_distributed_formation.py 文件中添加如下代码:

```python
#!/usr/bin/env python3
import rclpy
from rclpy.node import Node
from geometry_msgs.msg import Twist
from nav_msgs.msg import Odometry
from sensor_msgs.msg import LaserScan
from tf_transformations import euler_from_quaternion
from std_msgs.msg import Float32MultiArray
from std_msgs.msg import Int32MultiArray
import numpy as np
import math
import time

def quat_to_angle(quat):
    rot = euler_from_quaternion((quat.x, quat.y, quat.z, quat.w))
    return rot
```

```python
def normalize_angle(angle):
    res = angle
    while res > math.pi:
        res -= 2.0*math.pi
    while res < -math.pi:
        res += 2.0*math.pi
    return res

def proj(x, lb, ub):
    y = x.copy()
    y[x>ub] = ub
    y[x<lb] = lb
    return y

class OdomSubscriber(Node):
    def __init__(self):
        super().__init__('odom_subscriber')
        self.declare_parameter('tb3name', '')
        self.tb3name = self.get_parameter('tb3name').get_parameter_value().\
            string_value

        self_r1r2_list = ['robot1', 'robot2']
        self_r1r2_list.insert(0, self.tb3name) # 将机器人自己的名字加在首位

        self.odom_all = [None] * len(self_r1r2_list)
        for index, name in enumerate(self_r1r2_list):
            self.odom_monitor(index, name) # 订阅自己、robot1和robot2的里程计

    def odom_monitor(self, index, name):
        sub_odom = self.create_subscription(
            Odometry,
            f'/{name}/odom',
            lambda msg: self.odom_call_back(msg, index),
            10)

    def odom_call_back(self, msg, index):
        state_x = msg.pose.pose.position.x
        state_y = msg.pose.pose.position.y
        self.odom_all[index] = [state_x, state_y]

    def odom_sub_result(self):
```

```python
        return self.odom_all

class FindOptimal(Node):
    def __init__(self, odom, Delta12, S):
        super().__init__('find_optimal')
        self.declare_parameter('tb3name', '')
        self.declare_parameter('neighbor_name', '')
        self.declare_parameter('sigma', '')
        self.declare_parameter('upper_bound', '')
        self.tb3name = self.get_parameter('tb3name').get_parameter_value().
            string_value
        neighbor_name = self.get_parameter('neighbor_name').
            get_parameter_value().string_value
        sigma = self.get_parameter('sigma').get_parameter_value().string_value
        ub = self.get_parameter('upper_bound').get_parameter_value().
            string_value
        neighbor_list = neighbor_name.split()
        self.sigma = float(sigma)
        self.ub = float(ub)

        self.xi1 = np.array(odom[0]).reshape(-1, 1)  # 机器人1的初始位置常量\
            xi_1
        self.b3 = np.array(Delta12).reshape(-1, 1)  # 机器人1和机器人2的相对位置
            常量b_3

        self.p1 = np.array(odom[1]).reshape(-1, 1)  # 读取机器人1的里程计数据初
            始化优化变量p_1
        self.p12 = np.array(odom[2]).reshape(-1, 1)  # 读取机器人2的里程计数据初
            始化优化变量p_12
        self.y1 = np.zeros((2, 1))  # 辅助变量y_1初始值
        self.z1 = np.zeros((2, 1))  # 辅助变量z_1初始值
        self.w12 = np.zeros((2, 1))  # 辅助变量w_12初始值

        self.p1_neighbor = 2 * np.random.rand(len(neighbor_list), 2) - 1  # 初
            始化机器人邻居对p_1的估计
        self.p2_neighbor = 2 * np.random.rand(len(neighbor_list), 2) - 1  # 初
            始化机器人邻居对p_2的估计
        for index, name in enumerate(neighbor_list):
            self.p1p2_monitor(index, name)  # 初始化函数p1p2_monitor

        self.pub_p1p2 = self.create_publisher(  # 定义发布者用于发布机器人对p_1
            和p_2的估计值
```

```python
90              Float32MultiArray,
91              f'/{self.tb3name}/p1p2_est',
92              1)
93
94          self.timer = self.create_timer(0.05, self.p1p2_pub) # 优化算法求解p_1
                  和p_2的估计值并发布
95
96      def p1p2_monitor(self, index, name): # 订阅邻居对p1和p2的估计值
97          sub_p1p2 = self.create_subscription(
98              Float32MultiArray,
99              f'/{name}/p1p2_est',
100             lambda msg: self.p1p2_call_back(msg, index),
101             1)
102
103     def p1p2_call_back(self, msg, index):
104         self.p1_neighbor[index] = [msg.data[0], msg.data[1]]
105         self.p2_neighbor[index] = [msg.data[2], msg.data[3]]
106
107     def p1p2_pub(self):
108         p1_neighbor = np.array(self.p1_neighbor)
109         p2_neighbor = np.array(self.p2_neighbor)
110         p_1j = np.zeros((2,1))
111         p_2j = np.zeros((2,1))
112         for index in range(len(p1_neighbor)):
113             p_1j += self.p1-p1_neighbor[index].reshape(-1,1) # 自己与邻居对p_1
                      估计值差的累加
114             p_2j += self.p12-p2_neighbor[index].reshape(-1,1) # 自己与邻居对p_2
                      估计值差的累加
115
116         # 机器人1算法的迭代公式
117         p1 = proj(self.p1-self.sigma*(self.p1-self.xi1+self.y1+self.p1-self.
                  p12-self.b3+self.z1+p_1j), 0, self.ub)
118         p12 = proj(self.p12-self.sigma*(-self.y1-self.p1+self.p12+self.b3+self
                  .w12+p_2j), 0, self.ub)
119         y1 = self.y1+self.p1-self.p12-self.b3
120         z1 = self.z1+p_1j
121         w12 = self.w12+p_2j
122
123         self.p1 = p1
124         self.p12 = p12
125         self.y1 = y1
126         self.z1 = z1
```

```python
            self.w12 = w12
            current_p1p2 = np.array([self.p1, self.p12]).flatten()
            pub_p1p2_data = Float32MultiArray(data=current_p1p2)

            connections = self.pub_p1p2.get_subscription_count()
            if connections == len(p1_neighbor): # 如果所有的邻居连接成功，则发布当
                前的优化变量值
                self.pub_p1p2.publish(pub_p1p2_data)

    def opt_result(self):
        return self.tb3name, self.p1.flatten()

class RobotControl(Node):
    def __init__(self, goal):
        super().__init__('robot_control')
        self.declare_parameter('tb3name', 'turtlebot')
        tb3name = self.get_parameter('tb3name').get_parameter_value().\
            string_value

        self.scan_msg = None
        self.odom_msg = None
        self.is_arrival = False # 记录机器人是否到达目标位置
        self.goal = goal

        scan_sub = self.create_subscription(
            LaserScan,
            f'/{tb3name}/scan',
            self.scan_call_back,
            1)

        odom_sub = self.create_subscription(
            Odometry,
            f'/{tb3name}/odom',
            self.odom_call_back,
            1)

        self.robot_vel = self.create_publisher(
            Twist,
            f'/{tb3name}/cmd_vel',
            1)

        self.ini_for_scan()
```

```python
            self.ini_for_odom()
            robot_timer = self.create_timer(0.1, self.robot_control)

    def ini_for_scan(self):
        while rclpy.ok():
            if self.scan_msg is not None:
                self.get_logger().info('激光雷达完成初始化.')
                break
            rclpy.spin_once(self)
            time.sleep(0.01)

    def ini_for_odom(self):
        while rclpy.ok():
            if self.odom_msg is not None:
                self.get_logger().info('里程计完成初始化.')
                break
            rclpy.spin_once(self)
            time.sleep(0.01)

    def scan_call_back(self, msg):
        self.scan_msg = msg

    def odom_call_back(self, msg):
        self.odom_msg = msg

    def robot_control(self):
        goal_x = self.goal[0]
        goal_y = self.goal[1]
        scan_data = [self.scan_msg.ranges[0], self.scan_msg.ranges[20], self.
            scan_msg.ranges[340]]
        state_x = self.odom_msg.pose.pose.position.x
        state_y = self.odom_msg.pose.pose.position.y
        (roll, pitch, theta) = quat_to_angle(self.odom_msg.pose.pose.
            orientation)

        threshold = 0.4 # 激光雷达防碰撞的敏感值
        tolerance = 0.02 # 到达目标位置允许的误差值

        trans = (goal_x - state_x, goal_y - state_y)
        linear = math.sqrt(trans[0] ** 2 + trans[1] ** 2)
        angular = normalize_angle(math.atan2(trans[1], trans[0]) - theta)
        sign = lambda x: math.copysign(1, x) # 定义符号函数
```

```python
            cmd = Twist()

            if linear > tolerance: # 机器人移动和避障控制
                if scan_data[0] > threshold and scan_data[1] > threshold and
                    scan_data[2] > threshold:
                    if abs(angular) > math.pi/8:
                        cmd.linear.x = min(linear*0.8, 0.22)
                        cmd.angular.z = max(min(angular*0.8, 2.84), -2.84)
                    else:
                        cmd.linear.x = min(linear*1.0, 0.22)
                        cmd.angular.z = max(min(angular*0.1, 2.84), -2.84)
                else:
                    cmd.linear.x = -min(linear*0.1, 0.22)
                    cmd.angular.z = max(min(2.0 * sign(angular), 2.84), -2.84)
                    if scan_data[0] > threshold and scan_data[1] > threshold and
                        scan_data[2] > threshold:
                        cmd.linear.x = 0.1
                        cmd.angular.z = 0.0
            else:
                cmd.linear.x = 0.0
                cmd.angular.z = 0.0
                self.is_arrival = True
            self.robot_vel.publish(cmd)

    def arrival_result(self):
        return self.is_arrival

class MessageDiffusion(Node):
    def __init__(self):
        super().__init__('message_diffusion')
        self.declare_parameter('tb3name', '')
        self.declare_parameter('neighbor_name', '')
        tb3name = self.get_parameter('tb3name').get_parameter_value().
            string_value
        neighbor_name = self.get_parameter('neighbor_name').
            get_parameter_value().string_value
        self.neighbor_list = neighbor_name.split()
        robot_id = int(tb3name[-1])

        self.msg_pub = [robot_id] # 记录发布的信息
        self.pub_arriver = self.create_publisher( # 定义发布者用于发布到达目标
            位置的确认信息
```

```python
                Int32MultiArray,
                f'/{tb3name}/arriver_confirm',
                10)

        for name in self.neighbor_list:
            self.arriver_monitor(name)

        arriver_pub_timer = self.create_timer(0.1, self.arriver_pub) # 到达目
            标位置后发布确认信息

    def arriver_monitor(self, name): # 订阅邻居的确认信息
        sub_arriver = self.create_subscription(
            Int32MultiArray,
            f'/{name}/arriver_confirm',
            self.arriver_call_back,
            1)

    def arriver_call_back(self, msg):
        self.msg_pub = list(set(self.msg_pub+list(msg.data)))

    def arriver_pub(self):
        self.pub_arriver_data = Int32MultiArray(data=self.msg_pub)
        connections = self.pub_arriver.get_subscription_count()
        if connections == len(self.neighbor_list):
            self.pub_arriver.publish(self.pub_arriver_data)

    def arriver_result(self, robot_number):
        all_arrival = False
        if len(self.msg_pub) == robot_number: # 如果所有机器人都到达目标位置，
            则返回逻辑值True
            all_arrival = True
        return all_arrival

def main():
    rclpy.init()
    node = rclpy.create_node('node_ini')
    node.declare_parameter('robot_number', '')
    node.declare_parameter('max_time', '')
    robot_number = node.get_parameter('robot_number').get_parameter_value().
        string_value
    max_time = node.get_parameter('max_time').get_parameter_value().
        string_value
```

```python
robot_number = int(robot_number)
        max_time = float(max_time)
        node.destroy_node()

        S = np.zeros((3,robot_number,2))
        S[0] = np.array([[0,0],[2,0],[0,2],[2,2]])
        S[1] = np.array([[0,0],[1,0],[2,0],[1,2]])
        S[2] = np.array([[0,0],[2,0],[0,1],[2,1]])
        Delta12 = [[-1,0],[-0.5,0],[-1,-0.5]]

        for i in range(S.shape[0]):
            node = OdomSubscriber()
            while rclpy.ok():
                try:
                    rclpy.spin_once(node)
                    odom = node.odom_sub_result()
                except KeyboardInterrupt:
                    pass
                if None not in odom:
                    break
            node.destroy_node()

            node = FindOptimal(odom, Delta12[i], S[i])
            time_start, _ = node.get_clock().now().seconds_nanoseconds()
            while rclpy.ok():
                try:
                    rclpy.spin_once(node)
                    tb3name, goal = node.opt_result()
                except KeyboardInterrupt:
                    pass
                time_now, _ = node.get_clock().now().seconds_nanoseconds()
                if time_now - time_start > max_time:
                    node.get_logger().info('{0}得到的最优解：{1}'.format(tb3name,
                        goal))
                    break
            node.destroy_node()

            node = RobotControl(goal)
            while rclpy.ok():
                try:
                    rclpy.spin_once(node)
                    is_arrival = node.arrival_result()
```

```
323            except KeyboardInterrupt:
324                pass
325            if is_arrival:
326                break
327        node.destroy_node()
328
329        node = MessageDiffusion()
330        delay_counter = 0  # 延迟计数器
331        while rclpy.ok():
332            try:
333                rclpy.spin_once(node)
334                all_arrival = node.arriver_result(robot_number)
335            except KeyboardInterrupt:
336                pass
337            if all_arrival is True:
338                delay_counter += 1
339                if delay_counter == 20:
340                    node.get_logger().info('所有机器人都已到达目标位置.')
341                    break
342        node.destroy_node()
343    rclpy.shutdown()
344
345  if __name__ == '__main__':
346      main()
```

这里的代码与第 7.3.1 节相似，下面对其中不同的地方进行解释：

- 第 32~58 行定义了类 OdomSubscriber()，其内容与第 7.3.1 节定义的同名类的代码相似，主要用于订阅机器人自己、robot1 和 robot2 的里程计数据用于初始化分布式优化算法的变量;
- 第 60~136 行定义了类 FindOptimal()，是进行分布式优化计算的主要程序，包括以下内容：
    - 第 89~92 行定义了一个发布者，将 robot1 自身的变量 p_1 的值和对 robot2 的变量 p_2 的估计值以话题 "robot1/p1p2_est" 的形式发布出去;
    - 第 96~105 行定义了两个函数 p1p2_monitor() 和 p1p2_call_back() 用于订阅邻居对 p_1 和 p_2 的估计值;
    - 第 107~133 行定义了函数 p1p2_pub() 用于实现算法公式(3.25)计算最优解，其中第 117~121 行是具体实现公式(3.25)的代码。
- 第 138~230 行定义了类 RobotControl()，用于控制机器人移动到最优位置，与第 7.3.1 节的同名类代码相同;
- 第 232~273 行定义了类 MessageDiffusio()，用于确认是否所有的机器人都到达了目标位置，与第7.3.1节的同名类代码相同；

- 第 275~343 行定义了主函数，其中第 286~289 行给出了四个机器人的队形矩阵，其坐标对应图8.1中给出的三个队形，第 290 行给出了 robot1 和 robot2 之间的相对位置.

此外，对于 robot2，其代码与 robot1 类似，不同的地方是用下述的代码替换 robot1 代码的类 FindOptimal()，其他代码完全相同：

```
1   class FindOptimal(Node):
2       def __init__(self, odom, Delta12, S):
3           super().__init__('find_optimal')
4           self.declare_parameter('tb3name', '')
5           self.declare_parameter('neighbor_name', '')
6           self.declare_parameter('sigma', '')
7           self.declare_parameter('upper_bound', '')
8           self.tb3name = self.get_parameter('tb3name').get_parameter_value().
                string_value
9           neighbor_name = self.get_parameter('neighbor_name').
                get_parameter_value().string_value
10          sigma = self.get_parameter('sigma').get_parameter_value().string_value
11          ub = self.get_parameter('upper_bound').get_parameter_value().
                string_value
12          neighbor_list = neighbor_name.split()
13          self.sigma = float(sigma)
14          self.ub = float(ub)
15
16          self.xi2 = np.array(odom[0]).reshape(-1, 1)  # 机器人2的初始位置常量\
                xi_2
17          self.b3 = np.array(Delta12).reshape(-1, 1)  # 机器人1和机器人2的相对位置
                常量b_3
18
19          self.p2 = np.array(odom[2]).reshape(-1, 1)  # 读取机器人2的里程计数据初
                始化优化变量p_2
20          self.p21 = np.array(odom[1]).reshape(-1, 1)  # 读取机器人1的里程计数据初
                始化优化变量p_21
21          self.y2 = np.zeros((2, 1))  # 辅助变量y_2初始值
22          self.z21 = np.zeros((2, 1))  # 辅助变量z_21初始值
23          self.w2 = np.zeros((2, 1))  # 辅助变量w_2初始值
24
25          self.p1_neighbor = 2 * np.random.rand(len(neighbor_list), 2) - 1  # 初
                始化机器人邻居对p_1的估计
26          self.p2_neighbor = 2 * np.random.rand(len(neighbor_list), 2) - 1  # 初
                始化机器人邻居对p_2的估计
```

```python
            for index, name in enumerate(neighbor_list):
                self.p1p2_monitor(index, name) # 初始化函数p1p2_monitor

            self.pub_p1p2 = self.create_publisher( # 定义发布者用于发布机器人对p_1
                和p_2的估计值
                Float32MultiArray,
                f'/{self.tb3name}/p1p2_est',
                1)

            self.timer = self.create_timer(0.05, self.p1p2_pub) # 优化算法求解p_1
                和p_2的估计值并发布

        def p1p2_monitor(self, index, name): # 订阅邻居对p1和p2的估计值
            sub_p1p2 = self.create_subscription(
                Float32MultiArray,
                f'/{name}/p1p2_est',
                lambda msg: self.p1p2_call_back(msg, index),
                1)

        def p1p2_call_back(self, msg, index):
            self.p1_neighbor[index] = [msg.data[0], msg.data[1]]
            self.p2_neighbor[index] = [msg.data[2], msg.data[3]]

        def p1p2_pub(self):
            p1_neighbor = np.array(self.p1_neighbor)
            p2_neighbor = np.array(self.p2_neighbor)
            p_1j = np.zeros((2,1))
            p_2j = np.zeros((2,1))
            for index in range(len(p1_neighbor)):
                p_2j += self.p2-p2_neighbor[index].reshape(-1,1) # 自己与邻居对p_2
                    估计值差的累加
                p_1j += self.p21-p1_neighbor[index].reshape(-1,1) # 自己与邻居对p_1
                    估计值差的累加

            # 机器人2算法的迭代公式
            p2 = proj(self.p2-self.sigma*(self.p2-self.xi2+self.y2+self.p2-self.
                p21+self.b3+self.w2+p_2j), 0, self.ub)
            p21 = proj(self.p21-self.sigma*(-self.y2-self.p2+self.p21-self.b3+self
                .z21+p_1j), 0, self.ub)
            y2 = self.y2+self.p2-self.p21+self.b3
            z21 = self.z21+p_1j
            w2 = self.w2+p_2j
```

```
63
64            self.p2 = p2
65            self.p21 = p21
66            self.y2 = y2
67            self.z21 = z21
68            self.w2 = w2
69            current_p1p2 = np.array([self.p21, self.p2]).flatten()
70            pub_p1p2_data = Float32MultiArray(data=current_p1p2)
71
72            connections = self.pub_p1p2.get_subscription_count()
73            if connections == len(p1_neighbor): # 如果所有的邻居连接成功，则发布当
                   前的优化变量值
74                self.pub_p1p2.publish(pub_p1p2_data)
75
76        def opt_result(self):
77            return self.tb3name, self.p2.flatten()
```

类似地，对于 $robot\ i\ (i=3,4)$，其代码也与 robot1 类似，不同的地方是用下述的代码替换 robot1 代码的类 FindOptimal()，其他代码完全相同：

```
1   class FindOptimal(Node):
2       def __init__(self, odom, Delta12, S):
3           super().__init__('find_optimal')
4           self.declare_parameter('tb3name', '')
5           self.declare_parameter('neighbor_name', '')
6           self.declare_parameter('sigma', '')
7           self.declare_parameter('upper_bound', '')
8           self.tb3name = self.get_parameter('tb3name').get_parameter_value().
                string_value
9           neighbor_name = self.get_parameter('neighbor_name').
                get_parameter_value().string_value
10          sigma = self.get_parameter('sigma').get_parameter_value().string_value
11          ub = self.get_parameter('upper_bound').get_parameter_value().
                string_value
12          neighbor_list = neighbor_name.split()
13          self.sigma = float(sigma)
14          self.ub = float(ub)
15          robot_id = int(self.tb3name[-1]) # 取机器人名称的最后一个字符作为
                robot_id，即取值为3、4、5……
16
17          m, n = S.shape
18          self.M = np.array([[np.linalg.norm(S[1]), 0], [0, np.linalg.norm(S[1])
                ]])
```

```python
19          self.Mi = np.array([[S[robot_id-1][0], -S[robot_id-1][1]], [S[robot_id
                -1][1], S[robot_id-1][0]]])
20          self.Bi = np.hstack((self.M, self.Mi - self.M, - self.Mi))
21
22          self.xii = np.array(odom[0]).reshape(-1, 1) # 机器人i的初始位置常量\
                xi_i
23
24          self.pi = np.array(odom[0]).reshape(-1, 1) # 读取机器人i的里程计数据初
                始化优化变量p_i
25          self.pi1 = np.array(odom[1]).reshape(-1, 1) # 读取机器人1的里程计数据初
                始化优化变量p_i1
26          self.pi2 = np.array(odom[2]).reshape(-1, 1) # 读取机器人2的里程计数据初
                始化优化变量p_i2
27          self.yi = np.zeros((2, 1)) # 辅助变量y_i初始值
28          self.zi1 = np.zeros((2, 1)) # 辅助变量z_i1初始值
29          self.wi2 = np.zeros((2, 1)) # 辅助变量w_i2初始值
30
31          self.p1_neighbor = 2 * np.random.rand(len(neighbor_list), 2) - 1 # 初
                始化机器人邻居对p_1的估计
32          self.p2_neighbor = 2 * np.random.rand(len(neighbor_list), 2) - 1 # 初
                始化机器人邻居对p_2的估计
33          for index, name in enumerate(neighbor_list):
34              self.p1p2_monitor(index, name) # 初始化函数p1p2_monitor
35
36          self.pub_p1p2 = self.create_publisher( # 定义发布者用于发布机器人对p_1
                和p_2的估计值
37              Float32MultiArray,
38              f'/{self.tb3name}/p1p2_est',
39              1)
40
41          self.timer = self.create_timer(0.05, self.p1p2_pub) # 优化算法求解p_1
                和p2的估计值并发布
42
43      def p1p2_monitor(self, index, name): # 订阅邻居对p1和p2的估计值
44          sub_p1p2 = self.create_subscription(
45              Float32MultiArray,
46              f'/{name}/p1p2_est',
47              lambda msg: self.p1p2_call_back(msg, index),
48              1)
49
50      def p1p2_call_back(self, msg, index):
51          self.p1_neighbor[index] = [msg.data[0], msg.data[1]]
```

```python
            self.p2_neighbor[index] = [msg.data[2], msg.data[3]]

    def p1p2_pub(self):
        p1_neighbor = np.array(self.p1_neighbor)
        p2_neighbor = np.array(self.p2_neighbor)
        p_1j = np.zeros((2,1))
        p_2j = np.zeros((2,1))
        for index in range(len(p1_neighbor)):
            p_1j += self.pi1-p1_neighbor[index].reshape(-1,1) # 自己与邻居对p_1
                估计值差的累加
            p_2j += self.pi2-p2_neighbor[index].reshape(-1,1) # 自己与邻居对p_2
                估计值差的累加

        # 机器人i算法的迭代公式
        tpi = np.vstack((self.pi, self.pi1, self.pi2))
        pi = proj(self.pi-self.sigma*(self.pi-self.xii+self.M.dot(self.yi+self
            .Bi.dot(tpi))), 0, self.ub)
        pi1 = proj(self.pi1-self.sigma*((self.Mi.T-self.M).dot(self.yi+self.Bi
            .dot(tpi))+self.zi1+p_1j), 0, self.ub)
        pi2 = proj(self.pi2-self.sigma*(-self.Mi.T.dot(self.yi+self.Bi.dot(tpi
            ))+self.wi2+p_2j), 0, self.ub)
        yi = self.yi+self.Bi.dot(tpi)
        zi1 = self.zi1+p_1j
        wi2 = self.wi2+p_2j

        self.pi = pi
        self.pi1 = pi1
        self.pi2 = pi2
        self.yi = yi
        self.zi1 = zi1
        self.wi2 = wi2
        current_p1p2 = np.array([self.pi1, self.pi2]).flatten()
        pub_p1p2_data = Float32MultiArray(data=current_p1p2)

        connections = self.pub_p1p2.get_subscription_count()
        if connections == len(p1_neighbor): # 如果所有的邻居连接成功,则发布当
                前的优化变量值
            self.pub_p1p2.publish(pub_p1p2_data)

    def opt_result(self):
        return self.tb3name, self.pi.flatten()
```

上述 robot2 和 robot i 对应的代码分布用于实现算法公式(3.26)和(3.27)，由于其内容与 robot1 的代码类似，这里就不再详细解释。

打开 chapter8_2 目录下的 setup.py 文件，将 entry_points 域修改为如下形式：

```python
entry_points={
    'console_scripts': [
        'robot1_distributed_formation = chapter8_2.
            robot1_distributed_formation:main',
        'robot2_distributed_formation = chapter8_2.
            robot2_distributed_formation:main',
        'roboti_distributed_formation = chapter8_2.
            roboti_distributed_formation:main',
    ],
},
```

8.2.2 编写 launch 文件测试编队切换效果

在这一节中，编写一个 launch 文件测试上述队形切换代码的效果；在 chapter8_2 目录下新建一个 launch 目录，用以存放 launch 文件，运行如下命令：

```
$ cd ~/book_ws/src/chapter8_2
$ mkdir launch
```

在 launch 目录下新建一个名称为 tb3_distributed_formation_geo_demo.launch.py 的文件：

```
$ cd ~/book_ws/src/chapter8_2/launch
$ sudo nano tb3_distributed_formation_geo_demo.launch.py
```

并添加如下代码：

```python
#!/usr/bin/env python3
import os
from ament_index_python.packages import get_package_share_directory
from launch import LaunchDescription
from launch.actions import IncludeLaunchDescription
from launch.launch_description_sources import PythonLaunchDescriptionSource
from launch_ros.actions import Node

def generate_launch_description():
    TURTLEBOT3_MODEL = os.environ['TURTLEBOT3_MODEL']
    model_folder = 'turtlebot3_' + TURTLEBOT3_MODEL
    sdf_path = os.path.join(
```

```
13            get_package_share_directory('turtlebot3_gazebo'),
14            'models',
15            model_folder,
16            'model.sdf'
17        )
18
19        urdf_file_name = 'turtlebot3_' + TURTLEBOT3_MODEL + '.urdf'
20        urdf_path = os.path.join(
21            get_package_share_directory('turtlebot3_description'),
22            'urdf',
23            urdf_file_name)
24        with open(urdf_path, 'r') as infp:
25            robot_description = infp.read()
26
27        default_world_path = os.path.join(
28            get_package_share_directory('turtlebot3_gazebo'),
29            'worlds',
30            'empty_world.world'
31        )
32
33        robots = [
34            {'name': 'robot1', 'x_pose': '0.0', 'y_pose': '-1.0', 'z_pose': '0.01'
                , 'neighbor': 'robot3'},
35            {'name': 'robot2', 'x_pose': '0.0', 'y_pose': '0.0', 'z_pose': '0.01',
                'neighbor': 'robot3 robot4'},
36            {'name': 'robot3', 'x_pose': '0.0', 'y_pose': '1.0', 'z_pose': '0.01',
                'neighbor': 'robot1 robot2'},
37            {'name': 'robot4', 'x_pose': '0.0', 'y_pose': '2.0', 'z_pose': '0.01',
                'neighbor': 'robot2'},
38        ]
39
40        robot_number = str(len(robots)) # 机器人数量
41        max_time = '30' # 分布式优化算法最大运行时间
42        sigma = '0.05' # 优化算法步长\sigma
43        upper_bound = '2' # 编队移动范围的上界
44
45        gazebo_load_cmd = IncludeLaunchDescription(
46            PythonLaunchDescriptionSource(
47                os.path.join(get_package_share_directory('gazebo_ros'), 'launch',
                    'gazebo.launch.py')
48            ),
49            launch_arguments = {'world': default_world_path}.items()
```

```python
        )

        ld = LaunchDescription()
        ld.add_action(gazebo_load_cmd)

        for i in range(len(robots)):
            spawn_robot_cmd = Node(
                package = 'gazebo_ros',
                executable = 'spawn_entity.py',
                name = robots[i].get('name')+'_spawn_node',
                output = 'screen',
                arguments = [
                    '-entity', robots[i].get('name')+'_entity',
                    '-robot_namespace', robots[i].get('name'),
                    '-file', sdf_path,
                    '-x', robots[i].get('x_pose'),
                    '-y', robots[i].get('y_pose'),
                    '-z', robots[i].get('z_pose'),
                    ],
                )
            ld.add_action(spawn_robot_cmd)

            robot_state_publisher_cmd = Node(
                package = 'robot_state_publisher',
                executable = 'robot_state_publisher',
                namespace = robots[i].get('name'),
                name = robots[i].get('name')+'_state_node',
                output = 'screen',
                parameters = [{
                    'use_sim_time': True,
                    'robot_description': robot_description,
                    'publish_frequency': 50.0
                    }],
                )
            ld.add_action(robot_state_publisher_cmd)

        robot1_distributed_formation_cmd = Node(
            package = 'chapter8_2',
            executable = 'robot1_distributed_formation',
            name = robots[0].get('name')+'_node',
            output = 'screen',
            parameters = [{
```

```
            'tb3name': robots[0].get('name'),
            'neighbor_name': robots[0].get('neighbor'),
            'robot_number': robot_number,
            'max_time': max_time,
            'sigma': sigma,
            'upper_bound': upper_bound,
        }],
    )
    ld.add_action(robot1_distributed_formation_cmd)

    robot2_distributed_formation_cmd = Node(
        package = 'chapter8_2',
        executable = 'robot2_distributed_formation',
        name = robots[1].get('name')+'_node',
        output = 'screen',
        parameters = [{
            'tb3name': robots[1].get('name'),
            'neighbor_name': robots[1].get('neighbor'),
            'robot_number': robot_number,
            'max_time': max_time,
            'sigma': sigma,
            'upper_bound': upper_bound,
        }],
    )
    ld.add_action(robot2_distributed_formation_cmd)

    for i in range(2, len(robots)):
        roboti_distributed_formation_cmd = Node(
        package = 'chapter8_2',
        executable = 'roboti_distributed_formation',
        name = robots[i].get('name')+'_node',
        output = 'screen',
        parameters = [{
            'tb3name': robots[i].get('name'),
            'neighbor_name': robots[i].get('neighbor'),
            'robot_number': robot_number,
            'max_time': max_time,
            'sigma': sigma,
            'upper_bound': upper_bound,
        }],
    )
        ld.add_action(roboti_distributed_formation_cmd)
```

```
134
135         return ld
```

这里的代码与7.3.2相似, 对不同的地方解释如下:

- 第 33~38 行定义了变量 robots 用于存储四个机器人的信息, 包括机器人的名称、初始位置和邻居的名称;
- 第 40~43 行定义了一些常数, 包括机器人的数量 robot_number、分布式优化算法的最大计算时间 max_time、算法的步长 sigma 和机器人允许移动范围的上界 upper_bound, 这里我们在代码中将机器人允许移动范围的下界设置为零, 即投影函数的函数值区间是 [0, upper_bound];
- 第 87~101 行、第 103~117 行和第 119~134 行分别加载前面编写的 robot1, robot2 和 robot$i$ ($i = 3, 4$) 的节点文件, 实现对优化问题的分布式计算和对机器人的分布式控制.

重新打开 chapter8_2 目录下的 setup.py 文件, 将 data_files 域修改为如下形式:

```
1  data_files=[
2      ...
3      (os.path.join('share', package_name, 'launch'), glob(os.path.join('launch', '*.launch.py'))),
4  ],
```

并在文件头部添加如下代码:

```
1  import os
2  from glob import glob
```

接下来, 重新构建 chapter8_2 功能包, 运行如下命令:

```
$ cd ~/book_ws
$ colcon build --packages-select chapter8_2
```

打开终端, 输入如下命令进行测试:

```
$ ros2 launch chapter8_2 tb3_distributed_formation_geo_demo.launch.py
```

从图8.4给出的在 Gazebo 中的仿真结果可以看到四个机器人从直线队形出发, 依次切换成了图8.1所示的三个队形图标.

可以通过如下命令可视化节点和话题的关系图:

```
$ rqt_graph
```

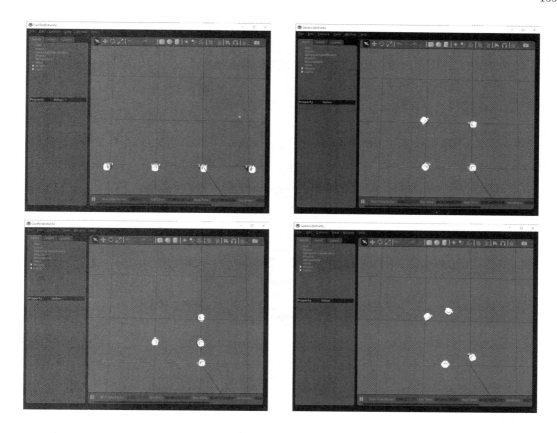

图 8.4　在 Gazebo 仿真中四个 TurtleBot 3 机器人的队形切换

如图8.5是截取的四个机器人进行分布式优化计算时的节点和话题关系图. 从图中可以看到四个机器人进行分布式优化计算时对话题 "p1p2_est" 的发布和订阅关系: 节点 "/robot1_node" 发布话题 "/robot1/p1p2_est", 同时订阅话题 "/robot3/p1p2_est", 这是由于 robot1 有一个邻居 robot3; 节点 "/robot2_node" 发布话题 "/robot2/p1p2_est", 同时订阅话题 "/robot3/p1p2_est" 和 "/robot4/p1p2_est", 这是由于 robot2 有两个邻居 robot3 和 robot4; 节点 "/robot3_node" 发布话题 "/robot3/p1p2_est", 同时订阅话题 "/robot1/p1p2_est" 和 "/robot2/p1p2_est", 这是由于 robot3 有两个邻居 robot1 和 robot2; 节点 "/robot4_node" 发布话题 "/robot4/p1p2_est", 同时订阅话题 "/robot2/p1p2_est", 这是由于 robot4 有一个邻居 robot2. 上述的通信关系明确反映了我们所设计的分布式计算框架, 即和图8.2的通信拓扑结构保持一致.

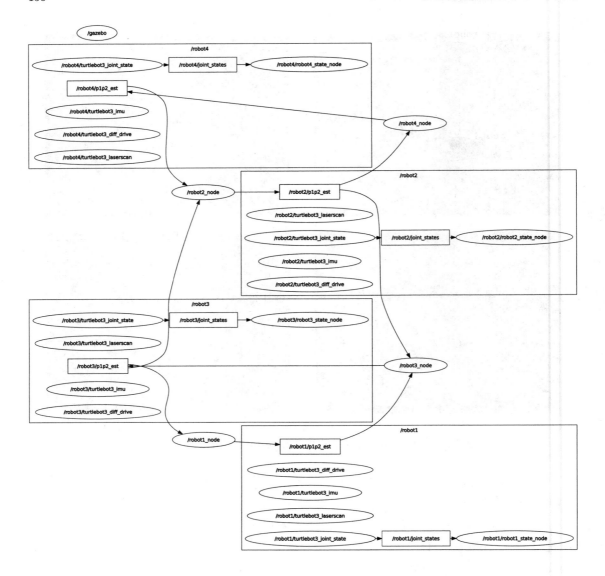

图 8.5　四个机器人进行分布式优化计算时的节点和话题关系图

⁂ 练　习 ⁂

1. 参考本章的代码, 对于第3章的集中式算法(3.15), 编写代码实现四个 TurtleBot 3 机器人的队形切换.
2. (切换拓扑下的多机器人编队)考虑如图8.6所示的两个通信拓扑结构, 假设机器人的通信网络在这两个拓扑结构下每隔 1 秒钟切换一次, 试参考本章的代码, 实现切换拓扑下的多机器人队形切换任务.

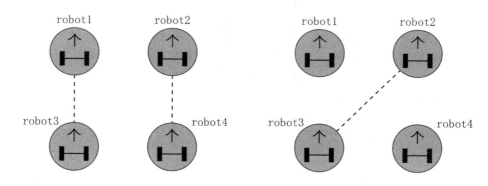

图 8.6　四个机器人形成的两个切换通信拓扑结构

第 9 章　基于时变分布式优化的多机器人围堵仿真

在前面两章中,实现的是多机器人编队问题的静态分布式优化算法. 静态优化算法的求解对机器人控制的实时性要求不是很高,所以在代码实现时机器人的优化求解和实时控制是相互独立的两个过程. 然而, 对于时变分布式优化问题, 这两个过程通常是统一的, 也就是机器人在实时控制的同时需要完成对时变优化问题的求解. 本章将基于前面第 4 章的时变分布式优化算法, 介绍相应的代码实现过程, 并在 ROS 2 的 Gazebo 仿真中实现多机器人的围堵任务.

9.1　时变分布式优化与实时求解算法

根据第 4 章中的描述, 假设有 4 个机器人来完成围堵任务. 被围堵目标 G 的移动轨迹 $c(t)$ 是一个与时间 t 相关的 2 维向量; 执行围堵任务的机器人 i 的空间坐标为 p_i; 机器人 i 的监测点坐标为 $s_i(t) = c(t) + (\cos((i-1)\pi/2), \sin((i-1)\pi/2))^\mathrm{T}$ $(i=1,2,3,4)$, 其坐标位置可以用来设置特定的队形. 图 9.1 展示了期望达到的围堵效果.

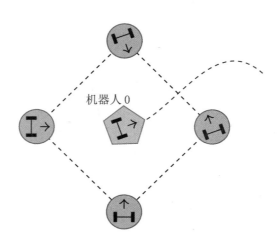

图 9.1　四个围堵机器人将机器人 0 围在了几何中心

根据第4章的公式(4.10), 实现上述围堵任务的优化问题可以写成

$$\min_{\boldsymbol{p}} \quad \frac{1}{2}\sum_{i=1}^{4}\left(\eta_i\|\boldsymbol{p}_i-\boldsymbol{s}_i(t)\|^2+(1-\eta_i)\|\boldsymbol{p}_i-\boldsymbol{c}(t)\|^2\right)$$
$$\text{s.t.} \quad \sum_{i=1}^{4}\boldsymbol{p}_i=4\boldsymbol{c}(t) \tag{9.1}$$

其中 $0\leqslant \eta_i \leqslant 1$ 是权重系数, 用于调节围堵机器人到围堵目标的距离与围堵机器人到监测点的距离的比例. 当 η_i 接近 0 时, 表示围堵机器人倾向于靠近围堵目标; 当 η_i 接近 1 时, 表示围堵机器人倾向于靠近监测点.

根据第4章的公式(4.19), 可以得到求解优化问题(9.1)的分布式优化算法为

$$\dot{\hat{\boldsymbol{p}}}_i = -\dot{\hat{\boldsymbol{\mu}}}_i - \sigma\boldsymbol{\nabla}_{\boldsymbol{p}_i}f_i(\hat{\boldsymbol{p}}_i,t) - \sigma\hat{\boldsymbol{\mu}}_i - \boldsymbol{\nabla}_{\boldsymbol{p}_i t}f_i(\hat{\boldsymbol{p}}_i,t) \tag{9.2a}$$
$$\dot{\hat{\boldsymbol{\mu}}}_i = -\sigma\boldsymbol{\nabla}_{\boldsymbol{p}_i}f_i(\hat{\boldsymbol{p}}_i,t) - \boldsymbol{\nabla}_{\boldsymbol{p}_i t}f_i(\hat{\boldsymbol{p}}_i,t) + \sigma(\hat{\boldsymbol{p}}_i-\boldsymbol{c}(t)) - \dot{\boldsymbol{c}}(t) - \sigma\hat{\boldsymbol{\mu}}_i$$
$$\quad -\alpha\sum_{j\in N_i}\text{sign}(\hat{\boldsymbol{\mu}}_i-\hat{\boldsymbol{\mu}}_j) \tag{9.2b}$$

其中 α 是正常数, $f_i(\hat{\boldsymbol{p}}_i,t)=(\eta_i\|\hat{\boldsymbol{p}}_i-\boldsymbol{s}_i(t)\|^2+(1-\eta_i)\|\hat{\boldsymbol{p}}_i-\boldsymbol{c}(t)\|^2)/2$, 因此

$$\boldsymbol{\nabla}_{\hat{\boldsymbol{p}}_i}f_i(\hat{\boldsymbol{p}}_i,t)=\eta_i(\hat{\boldsymbol{p}}_i-\boldsymbol{s}_i(t))+(1-\eta_i)(\hat{\boldsymbol{p}}_i-\boldsymbol{c}(t))$$
$$=\hat{\boldsymbol{p}}_i-\eta_i\boldsymbol{s}_i(t)-(1-\eta_i)\boldsymbol{c}(t)$$
$$\boldsymbol{\nabla}_{\hat{\boldsymbol{p}}_i t}f_i(\hat{\boldsymbol{p}}_i,t)=-\eta_i\dot{\boldsymbol{s}}_i(t)-(1-\eta_i)\dot{\boldsymbol{c}}(t)$$

9.2 分布式优化算法实现多机器人围堵任务

在这一节中, 将基于公式(9.2)编写代码实现四个 TurtleBot 3 机器人的围堵任务. 考虑图9.1中所示的四个机器人形成的正方形队形. 图中的虚线表示机器人之间的通信关系, 即四个机器人形成了环形的通信网络.

编写代码实现上述围堵任务, 算法流程如图9.2所示. 从该流程图中可以看出, 主要包括一个程序文件 tb3_distributed_containment.py. 围堵机器人实时获取被围堵目标的位置和速度信息, 同时订阅邻居的拉格朗日乘子数据, 通过分布式计算获得自己的目标位置, 然后控制自己向目标位置移动. 接下来, 将具体编写这个程序文件.

执行如下命令创建一个名称为 chapter9_2 的功能包:

```
$ cd ~/book_ws/src
$ ros2 pkg create --build-type ament_python chapter9_2
```

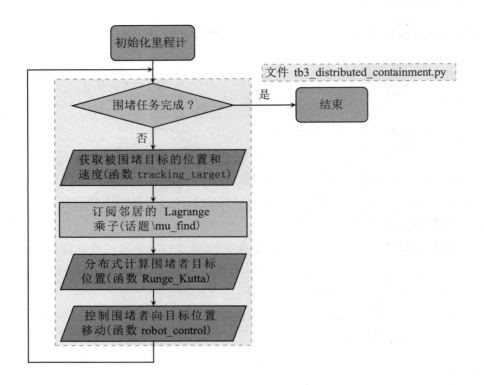

图 9.2　实现围堵任务的分布式算法流程图

然后构建这个包:

```
$ cd ~/book_ws
$ colcon build --packages-select chapter9_2
$ . install/setup.bash
```

9.2.1　实现机器人实时围堵控制

在 chapter9_2 功能包中创建一个实现机器人分布式实时围堵控制的 Python 节点文件:

```
$ cd ~/book_ws/src/chapter9_2/chapter9_2
$ sudo nano tb3_distributed_containment.py
```

在 tb3_distributed_containment.py 文件中添加如下代码:

```
1  #!/usr/bin/env python3
2  import rclpy
3  from rclpy.node import Node
4  from geometry_msgs.msg import Twist
5  from nav_msgs.msg import Odometry
6  from tf_transformations import euler_from_quaternion
```

```python
from std_msgs.msg import Float32MultiArray
import numpy as np
import math
import time

def quat_to_angle(quat):
    rot = euler_from_quaternion((quat.x, quat.y, quat.z, quat.w))
    return rot

def tracking_target(t): # 计算被围堵目标的位置和速度
    ct = 2 * np.array([np.cos(0.02 * t - 1/4*np.pi), np.sin(0.02 * t - 1/4*np
        .pi) + 1])
    dotct = 0.04 * np.array([-np.sin(0.02 * t - 1/4 * np.pi), np.cos(0.02 * t
        - 1/4 *np.pi)])
    return ct, dotct

def f(p_i, mu_i, t, robot_id, mu_neighbor): # 对应微分方程的右端项
    eta_i = 0.9
    sigma = 1
    alpha = 100
    ct, dotct = tracking_target(t) # 实时获取被围堵目标的位置和速度

    st_i = ct + np.array([np.cos((robot_id - 1)/2*np.pi), np.sin((robot_id -
        1)/2*np.pi)])
    dotst_i = dotct

    nabla_fp_i = p_i - eta_i * st_i - (1 - eta_i) * ct # 目标函数的梯度
    nabla_fpt_i = -eta_i * dotst_i - (1 - eta_i) * dotct # 目标函数的梯度关于t
        的导数

    mu_ij = np.zeros(2)
    for mu_j in mu_neighbor: # 累加mu_i-mu_j
        mu_ij += np.sign(mu_i - mu_j)

    dotmu_i = - sigma * nabla_fp_i - nabla_fpt_i + sigma * (p_i - ct) - dotct
        - sigma * mu_i - alpha * mu_ij
    dotp_i = - dotmu_i - sigma * nabla_fp_i - sigma * mu_i - nabla_fpt_i

    return dotp_i, dotmu_i

def Runge_Kutta(p_i, mu_i, t, robot_id, mu_neighbor): # 四阶Runge_Kutta算法
    step = 0.01
```

```python
44          k1_p, k1_mu = f(p_i, mu_i, t, robot_id, mu_neighbor)
45
46          t = t + step/2
47          p_temp = p_i
48          mu_temp = mu_i
49
50          p_i = p_temp + step/2 * k1_p
51          mu_i = mu_temp + step/2 * k1_mu
52          k2_p, k2_mu = f(p_i, mu_i, t, robot_id, mu_neighbor)
53
54          p_i = p_temp + step/2 * k2_p
55          mu_i = mu_temp + step/2 * k2_mu
56          k3_p, k3_mu = f(p_i, mu_i, t, robot_id, mu_neighbor)
57
58          t = t + step/2
59          p_i = p_temp + step * k3_p
60          mu_i = mu_temp + step * k3_mu
61          k4_p, k4_mu = f(p_i, mu_i, t, robot_id, mu_neighbor)
62
63          p_i = p_temp + step/6 * (k1_p + 2 * k2_p + 2 * k3_p + k4_p)
64          mu_i = mu_temp + step/6 * (k1_mu + 2 * k2_mu + 2 * k3_mu + k4_mu)
65          return p_i, mu_i
66
67      class RobotControl(Node):
68          def __init__(self):
69              super().__init__('robot_control')
70              self.declare_parameter('tb3name', '')
71              self.declare_parameter('neighbor_name', '')
72              tb3name = self.get_parameter('tb3name').get_parameter_value().
                      string_value
73              neighbor_name = self.get_parameter('neighbor_name').
                      get_parameter_value().string_value
74              neighbor_list = neighbor_name.split()
75
76              self.robot_id = int(tb3name[-1])
77              self.odom_msg = None
78              self.tb3pose = None
79              self.mu_i = 2 * np.random.rand(2) - 1  # 初始化机器人自己的Lagrange乘子
80
81              sub_odom = self.create_subscription(
82                  Odometry,
83                  f'/{tb3name}/odom',
```

```python
            self.odom_call_back,
            1)

        self.pub_mu = self.create_publisher(
            Float32MultiArray,
            f'/{tb3name}/mu_find',
            1)

        self.mu_neighbor = 2 * np.random.rand(len(neighbor_list), 2) - 1  # 初
            始化机器人邻居的Lagrange乘子
        for index, name in enumerate(neighbor_list):
            self.mu_monitor(index, name)  # 订阅邻居Lagrange乘子的值

        self.robot_vel = self.create_publisher(
            Twist,
            f'/{tb3name}/cmd_vel',
            1)

        self.ini_for_odom()
        second, nanosecond = self.get_clock().now().seconds_nanoseconds()  # 记
            录开始时间
        self.time_start = second + nanosecond/10**9  # 整个围堵任务开始的时间

    def ini_for_odom(self):
        while rclpy.ok():
            if self.odom_msg is not None:
                self.get_logger().info('里程计完成初始化.')
                break
            rclpy.spin_once(self)
            time.sleep(0.01)

    def odom_call_back(self, msg):
        self.odom_msg = msg
        state_x = msg.pose.pose.position.x
        state_y = msg.pose.pose.position.y
        (roll, pitch, theta) = quat_to_angle(msg.pose.pose.orientation)
        self.tb3pose = [state_x, state_y, theta]

    def mu_monitor(self, index, name):
        sub_mu = self.create_subscription(
            Float32MultiArray,
            f'/{name}/mu_find',
```

```python
                    lambda msg: self.mu_call_back(msg, index),
                    1)

    def mu_call_back(self, msg, index):
        self.mu_neighbor[index] = msg.data

    def robot_control(self):
        k_v = 1.0 # 控制系数
        k_e = 1.0 # 控制系数
        p_i = self.tb3pose[0:2]

        second, nanosecond = self.get_clock().now().seconds_nanoseconds() # 获
            得当前时间
        time_now = second + nanosecond/10**9
        t = time_now - self.time_start
        rk_t = time_now - time_now
        while rk_t < 0.02: # Runge_Kutta算法执行时间设置为0.02秒
            p_i, mu_i = Runge_Kutta(p_i, self.mu_i, t, self.robot_id, self.
                mu_neighbor)
            second, nanosecond = self.get_clock().now().seconds_nanoseconds()
            rk_now = second + nanosecond/10**9
            rk_t = rk_now - time_now
            t = rk_now - self.time_start
        goalp_i = p_i
        p_i = self.tb3pose[0:2]

        self.mu_i = mu_i
        pub_mu_data = Float32MultiArray(data=self.mu_i)

        connections = self.pub_mu.get_subscription_count()
        if connections == len(self.mu_neighbor):
            self.pub_mu.publish(pub_mu_data)

        deltap_i = goalp_i - p_i
        rho = np.linalg.norm(deltap_i, ord=2)
        theta = np.arctan2(deltap_i[1], deltap_i[0]) - self.tb3pose[2]

        cmd = Twist()
        v = k_v * np.cos(theta) * rho
        w = k_e * theta + k_v * np.sin(theta) * np.cos(theta)
        cmd.linear.x = max(min(v, 0.22), -0.22)
        cmd.angular.z = max(min(w, 2.84), -2.84)
```

```
164            self.robot_vel.publish(cmd)
165
166   def main():
167       rclpy.init()
168       node = RobotControl()
169       while rclpy.ok():
170           try:
171               rclpy.spin_once(node)
172               node.robot_control()
173           except KeyboardInterrupt:
174               pass
175       node.destroy_node()
176       rclpy.shutdown()
177
178   if __name__ == '__main__':
179       main()
```

下面解释上述代码的主要内容:

- 第 2~10 行加载了运行此文件需要用到的包, 和之前的代码用到的包基本相同;
- 第 16~19 行定义了函数 tracking_target(), 用于获取被围堵目标的位置和速度信息;
- 第 21~40 行定义了函数 $f()$, 用于计算公式(9.2)右端的表达式的值;
- 第 42~65 行定义了函数 Runge_Kutta(), 对微分方程(9.2)进行数值求解, 用于实时求解围堵机器人的目标移动位置;
- 第 67~164 行定义了类 RobotControl(), 主要包括以下内容:
  - 第 69~74 行对节点进行初始化, 并获取机器人自己和邻居的参数;
  - 第 87~90 行初始化一个发布者, 用于发布机器人自己的拉格朗日乘子的数值;
  - 第 92~94 行随机初始化机器人邻居的拉格朗日乘子, 并通过调用函数 mu_monitor() 获取邻居的拉格朗日乘子的数值;
  - 第 130~164 行定义了控制机器人移动的函数 robot_control(), 其中第 139~145 行通过调用 Runge_Kutta() 算法计算采样时间内机器人移动的目标位置, 以及拉格朗日乘子的数值, 并在第 153 行将此数值发布给邻居. 第 155~164 行基于附录D的机器人控制器设计方法, 计算机器人的线速度和角速度, 并将其发布给机器人.

打开 chapter9_2 目录下的 setup.py 文件, 将 entry_points 域修改为如下形式:

```
1   entry_points={
2       'console_scripts': [
3           'tb3_distributed_containment = chapter9_2.tb3_distributed_containment:
```

```
            main',
4       ],
5   },
```

9.2.2　编写 launch 文件测试实时围堵效果

在这一节中，编写一个 launch 文件测试上述实时围堵控制代码的效果；在 chapter9_2 目录下新建一个 launch 目录，用于存放 launch 文件，运行如下命令：

```
$ cd ~/book_ws/src/chapter9_2
$ mkdir launch
```

接下来，在 launch 目录下新建一个名称为 tb3_distributed_containment_demo.launch.py 的文件：

```
$ cd ~/book_ws/src/chapter9_2/launch
$ sudo nano tb3_distributed_containment_demo.launch.py
```

并添加如下代码：

```python
#!/usr/bin/env python3
import os
from ament_index_python.packages import get_package_share_directory
from launch import LaunchDescription
from launch.actions import IncludeLaunchDescription
from launch.launch_description_sources import PythonLaunchDescriptionSource
from launch_ros.actions import Node

def generate_launch_description():
    TURTLEBOT3_MODEL = os.environ['TURTLEBOT3_MODEL']
    model_folder = 'turtlebot3_' + TURTLEBOT3_MODEL
    sdf_path = os.path.join(
        get_package_share_directory('turtlebot3_gazebo'),
        'models',
        model_folder,
        'model.sdf'
        )
    
    urdf_file_name = 'turtlebot3_' + TURTLEBOT3_MODEL + '.urdf'
    urdf_path = os.path.join(
        get_package_share_directory('turtlebot3_description'),
        'urdf',
        urdf_file_name)
```

```python
with open(urdf_path, 'r') as infp:
    robot_description = infp.read()

default_world_path = os.path.join(
    get_package_share_directory('turtlebot3_gazebo'),
    'worlds',
    'empty_world.world'
    )

robots = [
    {'name': 'robot1', 'x_pose': '0.0', 'y_pose': '-1.0', 'z_pose': '0.01'
        , 'neighbor': 'robot2 robot4'},
    {'name': 'robot2', 'x_pose': '0.0', 'y_pose': '-2.0', 'z_pose': '0.01'
        , 'neighbor': 'robot1 robot3'},
    {'name': 'robot3', 'x_pose': '0.0', 'y_pose': '0.0', 'z_pose': '0.01',
        'neighbor': 'robot2 robot4'},
    {'name': 'robot4', 'x_pose': '0.0', 'y_pose': '1.0', 'z_pose': '0.01',
        'neighbor': 'robot1 robot3'}
    ]

gazebo_load_cmd = IncludeLaunchDescription(
    PythonLaunchDescriptionSource(
        os.path.join(get_package_share_directory('gazebo_ros'), 'launch',
            'gazebo.launch.py')
        ),
    launch_arguments = {'world': default_world_path}.items()
    )

ld = LaunchDescription()
ld.add_action(gazebo_load_cmd)

for i in range(len(robots)):
    spawn_robot_cmd = Node(
        package = 'gazebo_ros',
        executable = 'spawn_entity.py',
        name = robots[i].get('name')+'_spawn_node',
        output = 'screen',
        arguments = [
            '-entity', robots[i].get('name')+'_entity',
            '-robot_namespace', robots[i].get('name'),
            '-file', sdf_path,
            '-x', robots[i].get('x_pose'),
```

```python
                '-y', robots[i].get('y_pose'),
                '-z', robots[i].get('z_pose'),
            ],
        )
        ld.add_action(spawn_robot_cmd)

        robot_state_publisher_cmd = Node(
            package = 'robot_state_publisher',
            executable = 'robot_state_publisher',
            namespace = robots[i].get('name'),
            name = robots[i].get('name')+'_state_node',
            output = 'screen',
            parameters = [{
                'use_sim_time': True,
                'robot_description': robot_description,
                'publish_frequency': 50.0
            }],
        )
        ld.add_action(robot_state_publisher_cmd)

        tb3_formation_control_cmd = Node(
            package = 'chapter9_2',
            executable = 'tb3_distributed_containment',
            name = robots[i].get('name')+'_robot_control_node',
            output = 'screen',
            parameters = [{
                'tb3name': robots[i].get('name'),
                'neighbor_name': robots[i].get('neighbor'),
                'ini_x': robots[i].get('x_pose'),
                'ini_y': robots[i].get('y_pose')
            }],
        )
        ld.add_action(tb3_formation_control_cmd)

    return ld
```

这里的代码与第7.3.2节相似,不同的地方是第 82~94 行加载了前面编写的分布式实时围堵的程序,实现对优化问题的分布式计算和对围堵机器人的分布式控制.

重新打开 chapter9_2 目录下的 setup.py 文件,将 data_files 域修改为如下形式:

```python
data_files=[
    ...
```

```
3     (os.path.join('share', package_name, 'launch'), glob(os.path.join('launch
        ', '*.launch.py'))),
4   ],
```

并在文件头部添加如下代码:

```
1   import os
2   from glob import glob
```

接下来, 重新构建 chapter9_2 功能包, 运行如下命令:

```
$ cd ~/book_ws
$ colcon build --packages-select chapter9_2
```

打开终端, 输入如下命令进行测试:

```
$ ros2 launch chapter9_2 tb3_distributed_containment_demo.launch.py
```

从图9.3给出的在Gazebo中的仿真结果可以看到四个机器人分别从$(0,-2),(0,-1)$, $(0,0),(0,1)$出发, 逐渐形成正方形队形的合围之势.

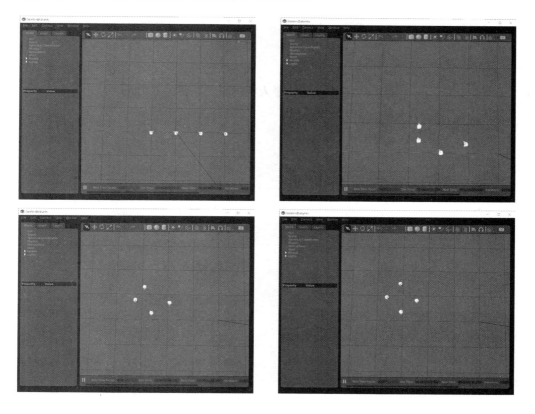

图 9.3　在 Gazebo 仿真中四个 TurtleBot 3 机器人逐渐形成正方形队形的合围之势

练 习

1. 参考第9.2节的代码,编写代码实现集中式优化算法(4.18).
2. 如图9.4所示,三个围堵机器人形成等边三角形队形将机器人 0 围在了几何中心,通过编写代码实现此围堵任务,要求:
    - 实时读取机器人 0 的里程计数据确定被围堵目标的位置;
    - 按照采样时间(如 0.1 秒)内机器人 0 的位置变化量确定被围堵机器人的速度.

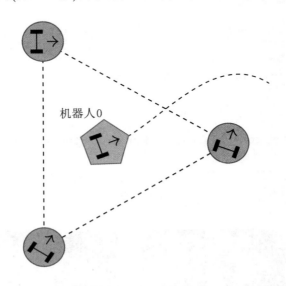

图 9.4 三个围堵机器人将机器人 0 围在了几何中心

第 3 部分

实 物 篇

第 10 章　基于分布式优化的多机器人实物编队

本章中,将在 TurtleBot 3 实物机器人上实现基于分布式优化的多机器人编队算法. 分两步来完成:第一步是实现单个机器人按照指定坐标点的定点移动, 第二步是编写优化算法实现多个机器人按照指定队形实现编队任务.

10.1　轮式里程计

为了实现机器人在物理空间的定点移动, 首先需要解决机器人在空间的定位问题. 机器人的定位方法有很多种, 有基于轮式里程计的, 有基于激光雷达的, 有基于相机的, 有基于 GPS 的, 有基于 UWB(超宽带, Ultra Wide Band) 的, 还有基于多种传感器融合的. 本书将不展开讨论机器人定位技术, 感兴趣的读者可以查阅相关书籍. 这里介绍基于轮式里程计的定位方法, 并在 TurtleBot 3 机器人上实现该方法.

通常所说的**定位**问题可以阐述为: 移动机器人根据自身状态和传感器数据实时确定自己在全局空间或者局部空间中的位置和姿态. 轮式里程计是一种较简单且成本较低的机器人定位方法. TurtleBot 3 机器人的两个车轮分别配有 Dynamixel 舵机, 可以实现基于轮式里程计的定位功能.

如图10.1所示, 以机器人左右轮的中间位置作为机器人本身在空间的位置, 因此机器人的移动线速度是机器人左右轮速度的平均值, 即 $v = (v_r + v_l)/2$.

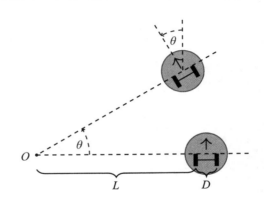

图 10.1　机器人车轮里程计导航推测

下面推导机器人的旋转角速度. 如图10.1所示, 已知机器人左右轮的速度分别是 v_l 和 v_r, 两个车轮的轴距记为 D. 假设机器人绕 O 点逆时针移动了一段很短的时间 Δt(在算法中通常作为采样时间), 绕 O 点旋转的角度为 θ, 机器人的姿态 (车头朝向) 变化量也是 θ. 从图中可以看出, 机器人左右轮行驶的路程分别为

$$\begin{cases} s_l = \theta L \\ s_r = \theta(L+D) \end{cases}$$

其中 L 是机器人左轮到 O 点的距离. 上式左右两边同时除以采样时间 Δt, 并令 $\Delta t \to 0$, 可以得到

$$\begin{cases} v_l = \omega L \\ v_r = \omega(L+D) \end{cases}$$

其中 ω 是机器人的旋转角速度. 由此可以推出

$$\omega = \frac{v_r - v_l}{D}$$

即机器人的旋转角速度等于机器人的右轮与左轮速度差除以车轮轴距.

有了机器人的线速度和角速度, 便可以估计出机器人在空间的相对位置坐标和姿态. 假设机器人的移动采样时间仍然是 Δt, 那么可以利用如下公式实时推演机器人的位置和姿态:

$$\begin{cases} x_{t+1} = x_t + v_t \cos(\theta_t) \Delta t \\ y_{t+1} = y_t + v_t \sin(\theta_t) \Delta t \\ \theta_{t+1} = \theta_t + \omega_t \Delta t \end{cases} \tag{10.1}$$

其中 x_t, y_t, θ_t, v_t 和 ω_t 分别是机器人在 t 时刻的 x 坐标、y 坐标、姿态、线速度和角速度.

下面根据上述的轮式里程计原理编写代码实现对 TurtleBot 3 机器人位置坐标的估计, 并控制机器人向指定点移动.

通过远程连接方式登录 TurtleBot 3 的 Ubuntu 系统[1]. 这里推荐使用 Visual Studio Code (简称 VS Code) 软件, 并安装 Remote SSH 插件远程登录 TurtleBot 3 的 Ubuntu 系统, 这样就可以实现在本地的 VS Code 软件中对远程代码进行编辑和调试.

通过 VS Code 远程连接到 TurtleBot 3 后, 新建一个终端窗口, 通过如下命令创建工作目录并设置环境:

```
$ mkdir -p ~/book_tb3/src
$ cd ~/book_tb3
$ colcon build
```

---

[1] 本书使用的 TurtleBot 3 实物安装的是树莓派 4B 开发板, 其操作系统和 ROS 2 的安装可以参考官方教程关于 Foxy 的安装步骤(https://emanual.robotis.com/docs/en/platform/turtlebot3/sbc_setup).

```
$ echo "source ~/book_tb3/install/setup.bash" >> ~/.bashrc
$ source ~/.bashrc
```

然后执行如下命令创建一个名称为 chapter10_1_tb3 的功能包：

```
$ cd ~/book_tb3/src
$ ros2 pkg create --build-type ament_python chapter10_1_tb3
```

构建这个包：

```
$ cd ~/book_tb3
$ colcon build --packages-select chapter10_1_tb3
$ . install/setup.bash
```

10.1.1 利用轮式里程计实现机器人定位

在 chapter10_1_tb3 功能包中创建一个利用轮式里程计实现机器人定位的 Python 节点文件：

```
$ cd ~/book_tb3/src/chapter10_1_tb3/chapter10_1_tb3
$ sudo nano tb3_wheel_odom.py
```

在 tb3_wheel_odom.py 文件中添加如下代码：

```python
#!/usr/bin/env python3
import rclpy
from rclpy.node import Node
from nav_msgs.msg import Odometry
from sensor_msgs.msg import JointState
from tf_transformations import quaternion_from_euler
import math

def normalize_angle(angle):
    res = angle
    while res > math.pi:
        res -= 2.0*math.pi
    while res < -math.pi:
        res += 2.0*math.pi
    return res

class OdomPublish(Node):
    def __init__(self):
        super().__init__('odom_publish')
```

```
21         self.odom_pub = self.create_publisher(
22             Odometry,
23             'odom_tb3',
24             1)
25
26         joint_sub = self.create_subscription(
27             JointState,
28             "joint_states",
29             self.joint_call_back, # 回调函数订阅左右车轮的线速度
30             1)
31
32         self.wheel_left = 0.0 # 左车轮初始线速度
33         self.wheel_right = 0.0 # 右车轮初始线速度
34         self.wheel_base = 0.177 # 车轮轴距
35         self.state_x = 0.0 # 机器人初始位置坐标是(0.0, 0.0)
36         self.state_y = 0.0
37         self.state_theta = 0.0 # 机器人初始姿态角是0.0
38
39         second, nanosecond = self.get_clock().now().seconds_nanoseconds() # 记
                录开始时间
40         self.current_time = second + nanosecond/10**9
41
42         odom_timer = self.create_timer(0.01, self.odom_publish) # 以周期0.01秒
                发布里程计
43
44     def joint_call_back(self, msg):
45         velocity_msg = msg.velocity
46         self.wheel_left = velocity_msg[0]
47         self.wheel_right = velocity_msg[1]
48
49     def odom_publish(self):
50         prev_update_time = self.current_time
51         odom = Odometry()
52         odom.header.stamp = self.get_clock().now().to_msg()
53         odom.header.frame_id = 'odom_tb3'
54         odom.child_frame_id = 'base_footprint'
55
56         second, nanosecond = self.get_clock().now().seconds_nanoseconds() # 重
                新获取当前时间
57         self.current_time = second + nanosecond/10**9
58         delta_t = self.current_time - prev_update_time # 计算前后两次采样时间间
                隔
```

```python
            v = (self.wheel_right + self.wheel_left) / 2.0  # 机器人的移动线速度
            omega = (self.wheel_right - self.wheel_left) / self.wheel_base  # 机器人的旋转角速度

            # 实时推演机器人相对初始点的位置和姿态
            self.state_x += v * math.cos(self.state_theta) * delta_t
            self.state_y += v * math.sin(self.state_theta) * delta_t
            self.state_theta = normalize_angle(self.state_theta + omega * delta_t)

            odom.pose.pose.position.x = self.state_x
            odom.pose.pose.position.y = self.state_y
            odom.pose.pose.position.z = 0.0
            quat = quaternion_from_euler(0.0, 0.0, self.state_theta)
            odom.pose.pose.orientation.x = quat[0]
            odom.pose.pose.orientation.y = quat[1]
            odom.pose.pose.orientation.z = quat[2]
            odom.pose.pose.orientation.w = quat[3]
            odom.twist.twist.linear.x = v
            odom.twist.twist.linear.y = 0.0
            odom.twist.twist.linear.z = 0.0
            odom.twist.twist.angular.x = 0.0
            odom.twist.twist.angular.y = 0.0
            odom.twist.twist.angular.z = omega
            self.odom_pub.publish(odom)

def main():
    rclpy.init()

    node = OdomPublish()
    try:
        rclpy.spin(node)
    except KeyboardInterrupt:
        pass
    node.destroy_node()
    rclpy.shutdown()

if __name__ == '__main__':
    main()
```

下面解释上述代码的主要内容：

- 第 2~7 行加载了运行此文件需要用到的包，其中第 5 行加载了 ROS 的传感器消

息类型的 JointState 接口, 用于订阅 TurtleBot 3 左右车轮的线速度;
- 第 17~82 行定义了类 OdomPublish(), 主要包括以下内容:
    - 第 21~24 行创建了一个发布者, 用于发布机器人的里程计数据, 发布的话题名称是 odom_tb3;
    - 第 26~30 行创建了一个订阅者, 用于订阅机器人左右车轮的线速度, 并通过调用回调函数 joint_call_back() 获取车轮的速度信息;
    - 第 32~37 行是一些参数的初始化, 其含义在代码中进行了解释;
    - 第 42 行启动一个计时器, 按照周期 0.01 秒调用回调函数 odom_publish() 用于发布机器人的里程计数据;
    - 第 44~47 行定义了回调函数 joint_call_back(), 用于获取机器人左右轮的线速度信息, 并将其存储在全局变量中;
    - 第 49~82 行定义了回调函数 odom_publish(), 基于上节的轮式里程计原理计算机器人的位姿, 并将其发布出去. 其中第 60~61 行用于计算机器人的线速度和角速度, 第 64~66 行根据公式(10.1)累计计算机器人的位姿, 第 68~82 行将机器人的位姿数据通过 Odometry 接口存储在变量 odom 中, 并将其以话题 odom_tb3 的形式发布出去.
- 第 84~93 行是主程序, 和之前的代码相似.

10.1.2 编写定点移动代码

下面基于上一节的轮式里程计定位数据实现 TurtleBot 3 机器人按照指定坐标点移动. 通过如下命令创建一个实现定点移动的 Python 节点文件:

```
$ cd ~/book_tb3/src/chapter10_1_tb3/chapter10_1_tb3
$ sudo nano tb3_move_to_goal.py
```

在 tb3_move_to_goal.py 文件中添加如下代码:

```
1  #!/usr/bin/env python3
2  import rclpy
3  from rclpy.node import Node
4  from nav_msgs.msg import Odometry
5  from geometry_msgs.msg import Twist
6  from tf_transformations import euler_from_quaternion
7  import math
8  import time
9
10 def quat_to_angle(quat):
11     rot = euler_from_quaternion((quat.x, quat.y, quat.z, quat.w))
12     return rot
13
```

```python
14  def normalize_angle(angle):
15      res = angle
16      while res > math.pi:
17          res -= 2.0*math.pi
18      while res < -math.pi:
19          res += 2.0*math.pi
20      return res
21
22  class RobotControl(Node):
23      def __init__(self):
24          super().__init__('robot_control')
25          self.odom_msg = None
26          self.is_arrival = False  # 记录机器人是否到达目标位置
27
28          odom_sub = self.create_subscription(
29              Odometry,
30              'odom_tb3',
31              self.odom_call_back,
32              1)
33
34          self.robot_vel = self.create_publisher(
35              Twist,
36              'cmd_vel',
37              1)
38
39      def odom_call_back(self, msg):
40          self.odom_msg = msg
41
42      def robot_control(self, goal):
43          k_v = 1.0  # 控制系数
44          k_e = 1.0  # 控制系数
45          goal_x = goal[0]
46          goal_y = goal[1]
47          state_x = self.odom_msg.pose.pose.position.x
48          state_y = self.odom_msg.pose.pose.position.y
49          (roll, pitch, theta) = quat_to_angle(self.odom_msg.pose.pose.
                  orientation)
50
51          tolerance = 0.01  # 到达目标位置允许的误差值
52
53          trans = (goal_x - state_x, goal_y - state_y)
54          linear = math.sqrt(trans[0] ** 2 + trans[1] ** 2)
```

```python
            angular = normalize_angle(math.atan2(trans[1], trans[0]) - theta)
            cmd = Twist()

            if linear > tolerance:
                v = k_v * math.cos(angular) * linear
                w = k_e * angular + k_v * math.sin(angular) * math.cos(angular)
                cmd.linear.x = max(min(v, 0.12), -0.12)
                cmd.angular.z = max(min(w, 0.5), -0.5)
            else:
                cmd.linear.x = 0.0
                cmd.angular.z = 0.0
                self.is_arrival = True
            self.robot_vel.publish(cmd)

    def arrival_result(self):
        return self.is_arrival

def main():
    rclpy.init()
    goals = [[0.0, 1.0], [1.0, 1.0], [0.0, 2.0]]

    for goal in goals:
        node = RobotControl()
        node.get_logger().info('\n 机器人目标位置: ' + str(goal))
        while rclpy.ok():
            try:
                rclpy.spin_once(node)
                node.robot_control(goal)
                is_arrival = node.arrival_result()
            except KeyboardInterrupt:
                pass
            if is_arrival == True:
                break
        node.destroy_node()
        time.sleep(1.0)
    rclpy.shutdown()

if __name__ == '__main__':
    main()
```

这里的代码与第9.2.1节关于定点移动的代码相似,这里不再赘述.

打开 chapter10_1_tb3 目录下的 setup.py 文件,将 entry_points 域修改为如下

形式:

```
entry_points={
    'console_scripts': [
        'tb3_wheel_odom = chapter10_1_tb3.tb3_wheel_odom:main',
        'tb3_move_to_goal = chapter10_1_tb3.tb3_move_to_goal:main',
    ],
},
```

10.1.3 编写 launch 文件测试定点移动效果

在这一节中，编写一个 launch 文件测试上述定点移动代码的效果；在 chapter10_1_tb3 目录下新建一个 launch 目录，用于存放 launch 文件，运行如下命令:

```
$ cd ~/book_tb3/src/chapter10_1_tb3
$ mkdir launch
```

在 launch 目录下新建一个名称为 tb3_move_to_goal_demo.launch.py 的文件:

```
$ cd ~/book_tb3/src/chapter10_1_tb3/launch
$ sudo nano tb3_move_to_goal_demo.launch.py
```

并添加如下代码:

```
#!/usr/bin/env python3
import os
from ament_index_python.packages import get_package_share_directory
from launch import LaunchDescription
from launch_ros.actions import Node

def generate_launch_description():
    TURTLEBOT3_MODEL = os.environ['TURTLEBOT3_MODEL']

    tb3_param_file = os.path.join(
        get_package_share_directory('turtlebot3_bringup'),
        'param',
        TURTLEBOT3_MODEL + '.yaml')

    urdf_file_name = 'turtlebot3_' + TURTLEBOT3_MODEL + '.urdf'
    urdf_path = os.path.join(
        get_package_share_directory('turtlebot3_description'),
        'urdf',
        urdf_file_name)
    with open(urdf_path, 'r') as infp:
```

```python
        robot_description = infp.read()

    ld = LaunchDescription()

    robot_state_publisher_cmd = Node(
        package = 'robot_state_publisher',
        executable = 'robot_state_publisher',
        name = 'tb3_state_publisher',
        output = 'screen',
        parameters = [{
            'use_sim_time': False,
            'robot_description': robot_description,
            'publish_frequency': 50.0
        }]
    )
    ld.add_action(robot_state_publisher_cmd)

    hlds_laser_publisher_cmd = Node(
        package = 'hls_lfcd_lds_driver',
        executable = 'hlds_laser_publisher',
        name = 'tb3_hlds_laser_publisher',
        output='screen',
        parameters = [{
            'port': '/dev/ttyUSB0',
            'frame_id': 'base_scan'
        }]
    )
    ld.add_action(hlds_laser_publisher_cmd)

    tb3_node_cmd = Node(
        package = 'turtlebot3_node',
        executable = 'turtlebot3_ros',
        output='screen',
        parameters = [tb3_param_file],
        arguments = ['-i', '/dev/ttyACM0']
    )
    ld.add_action(tb3_node_cmd)

    tb3_odom_pub_cmd = Node(
        package = 'chapter10_1_tb3',
        executable = 'tb3_wheel_odom',
        name = 'tb3_odom_node',
```

```
63          output = 'screen'
64      )
65      ld.add_action(tb3_odom_pub_cmd)
66  
67      tb3_move_to_goal_cmd = Node(
68          package = 'chapter10_1_tb3',
69          executable = 'tb3_move_to_goal',
70          name = 'tb3_move_node',
71          output = 'screen'
72      )
73      ld.add_action(tb3_move_to_goal_cmd)
74  
75      return ld
```

上述 launch 文件的代码与前述仿真中的代码相似，但是由于是基于实物实现的，其内容与仿真部分又有许多不同，下面对代码中不同的地方进行解释：

- 第 2~5 行加载了运行此 launch 文件需要用到的包；
- 第 10~13 行定义了参数文件，用于第 54 行加载 turtlebot3_ros 模块时的参数项；
- 第 38~48 行通过运行 hls_lfcd_lds_driver 包的 hlds_laser_publisher 节点文件加载机器人激光雷达传感器；
- 第 50~57 行通过运行 turtlebot3_node 包的 turtlebot3_ros 节点文件加载机器人底盘 OpenCR 控制器；
- 第 59~65 行和第 67~73 行分别加载前面编写的发布机器人里程计和控制机器人定点移动的节点文件。

重新打开 chapter10_1_tb3 目录下的 setup.py 文件，将 data_files 域修改为如下形式：

```
1  data_files=[
2      ...
3      (os.path.join('share', package_name, 'launch'), glob(os.path.join('launch
       ', '*.launch.py'))),
4  ],
```

并在文件头部添加如下代码：

```
1  import os
2  from glob import glob
```

接下来，重新构建 chapter10_1_tb3 功能包，运行如下命令：

```
$ cd ~/book_tb3
$ colcon build --packages-select chapter10_1_tb3
```

打开终端，输入如下命令进行测试：

```
$ ros2 launch chapter10_1_tb3 tb3_move_to_goal_demo.launch.py
```

将机器人的移动范围限制在边长是 2 米的正方形区域内。从图10.2给出的 TurtleBot 3 机器人的实物运行结果可以看到机器人从(0,0)点出发，首先移动到(0,1)点，再到(1,1)点，最后移动到(0,2)点。由于只使用里程计对机器人进行定位，通常定位的累积误差会比较大。从实物的实验结果也可以看出，机器人在移动到前两个目标位置时，累积误差比较小，但是对于第三个目标位置，其累积误差已经比较大了。

图 10.2　在实际环境中 TurtleBot 3 机器人的定点移动

10.2　ROS 2 导航工具包 Nav 2

从上一节的轮式里程计的实验结果可以看出，机器人仅使用车轮编码器进行定位，其累积误差往往会比较大，因此需要寻求更精准的定位方法。ROS 2 的导航工具包 Nav 2 (Navigation 2)[①]可以帮助实现机器人的定位和导航，有关该工具包的介绍，读者可以查阅文献 [11]。

通过如下命令在 TurtleBot 3 上安装 Nav 2 导航包：

```
$ sudo apt install ros-humble-navigation2
$ sudo apt install ros-humble-nav2-bringup
```

---

① 有关 Navigation 2 的介绍可以参考官方文档(https://navigation.ros.org)。

下面利用 Nav 2 导航包编写代码控制 TurtleBot 3 机器人进行定点移动. 同样地, 使用 VS Code 软件远程登录 TurtleBot 3 的 Ubuntu 系统. 新建一个终端窗口, 执行如下命令创建一个名称为 chapter10_2_tb3 的功能包:

```
$ cd ~/book_tb3/src
$ ros2 pkg create --build-type ament_python chapter10_2_tb3
```

构建这个包:

```
$ cd ~/book_tb3
$ colcon build --packages-select chapter10_2_tb3
$ . install/setup.bash
```

10.2.1 利用 Nav 2 编写定点移动代码

利用 ROS 2 的 Nav 2 导航包编写代码实现 TurtleBot 3 机器人按照指定坐标点移动. 通过如下命令创建一个实现定点移动的 Python 节点文件:

```
$ cd ~/book_tb3/src/chapter10_2_tb3/chapter10_2_tb3
$ sudo nano nav2_move_to_goal.py
```

在 nav2_move_to_goal.py 文件中添加如下代码:

```python
#!/usr/bin/env python3
import rclpy
from rclpy.node import Node
from geometry_msgs.msg import PoseStamped
from nav2_simple_commander.robot_navigator import BasicNavigator
import time


def main():
    rclpy.init()
    navigator = BasicNavigator()
    navigator.declare_parameter('ini_x', '')
    navigator.declare_parameter('ini_y', '')
    ini_x = navigator.get_parameter('ini_x').get_parameter_value().\
        string_value
    ini_y = navigator.get_parameter('ini_y').get_parameter_value().\
        string_value
    ini_x = float(ini_x)
    ini_y = float(ini_y)

    # 设置机器人的初始位姿
    initial_pose = PoseStamped()
```

```
20          initial_pose.header.frame_id = 'map'
21          initial_pose.header.stamp = navigator.get_clock().now().to_msg()
22          initial_pose.pose.position.x = ini_x
23          initial_pose.pose.position.y = ini_y
24          navigator.setInitialPose(initial_pose)
25
26          navigator.waitUntilNav2Active() # 等待激活Nav2
27
28          goals = [[0.8, 0.0], [0.8, 0.8], [0.0, 0.0]] # 机器人需要到达的目标位置列
                表
29          for goal in goals:
30              goal_x = goal[0]
31              goal_y = goal[1]
32              navigator.get_logger().info('\n 机器人目标位置: ' + str(goal))
33
34              # 移动机器人到指定目标位置
35              goal_pose = PoseStamped()
36              goal_pose.header.frame_id = 'map'
37              goal_pose.header.stamp = navigator.get_clock().now().to_msg()
38              goal_pose.pose.position.x = goal_x
39              goal_pose.pose.position.y = goal_y
40              time.sleep(1.0)
41              navigator.goToPose(goal_pose)
42
43              i = 0
44              while not navigator.isTaskComplete():
45                  i += 1
46                  if i % 10 == 0:
47                      navigator.get_logger().info('\n 机器人正在前往目标位置: ' + str(
                            goal))
48
49      if __name__ == '__main__':
50          main()
```

下面解释上述代码的主要内容:

- 第 2~6 行加载了运行此文件需要用到的包,其中第 4 行加载了带有时间戳的机器人姿态消息类型接口,用于设置机器人的初始位姿和目标位姿,第 5 行加载了 Nav 2 的 BasicNavigator 包,用于实现机器人的定点导航;
- 第 8~47 行是主程序,主要包括以下内容:
    - 第 10 行实例化类 BasicNavigator(),后续的代码主要是对这个实例进行设置,并使用 Nav 2 的导航功能实现机器人的定点移动;

- 第 11~16 行获取机器人的初始坐标位置信息;
- 第 18~24 行设置机器人的初始位姿,并在第 24 行使用 setInitialPose 方法初始化机器人的位姿;
- 第 26 行等待激活 Nav 2 功能;
- 第 28 行设置机器人需要到达的坐标位置,这里我们设置了三个坐标位置,希望机器人依次移动到这三个位置;
- 第 29~47 行利用循环结构设置机器人的目标位置,其中第 30~31 行定义了机器人目标位置的坐标,第 35~41 行设置了目标位置并使用 goToPose 方法实现机器人向目标位置移动,第 43~47 行用于在机器人移动过程中显示一些提示信息(读者可以修改这里的代码显示或者实现其它效果).

打开 chapter10_2_tb3 目录下的 setup.py 文件,将 entry_points 域修改为如下形式:

```
entry_points={
    'console_scripts': [
        'nav2_move_to_goal = chapter10_2_tb3.nav2_move_to_goal:main',
    ],
},
```

10.2.2 编写 launch 文件测试定点移动效果

在这一节中,编写一个 launch 文件测试上述定点移动代码的效果;在 chapter10_2_tb3 目录下新建一个 launch 目录,用于存放 launch 文件,运行如下命令:

```
$ cd ~/book_tb3/src/chapter10_2_tb3
$ mkdir launch
```

在 launch 目录下新建一个名称为 nav2_move_to_goal_demo.launch.py 的文件:

```
$ cd ~/book_tb3/src/chapter10_2_tb3/launch
$ sudo nano nav2_move_to_goal_demo.launch.py
```

并添加如下代码:

```
#!/usr/bin/env python3
import os
from ament_index_python.packages import get_package_share_directory
from launch import LaunchDescription
from launch.actions import IncludeLaunchDescription
from launch.launch_description_sources import PythonLaunchDescriptionSource
from launch_ros.actions import Node

```

```python
def generate_launch_description():
    TURTLEBOT3_MODEL = os.environ['TURTLEBOT3_MODEL']

    tb3_param_file = os.path.join(
        get_package_share_directory('turtlebot3_bringup'),
        'param',
        TURTLEBOT3_MODEL + '.yaml')

    urdf_file_name = 'turtlebot3_' + TURTLEBOT3_MODEL + '.urdf'
    urdf_path = os.path.join(
        get_package_share_directory('turtlebot3_description'),
        'urdf',
        urdf_file_name)
    with open(urdf_path, 'r') as infp:
        robot_description = infp.read()

    map_file = os.path.join(
        get_package_share_directory('chapter10_2_tb3'),
        'map',
        'blank_map.yaml')

    nav2_params_file = os.path.join(
        get_package_share_directory('turtlebot3_navigation2'),
        'param',
        'burger.yaml')

    ld = LaunchDescription()

    nav2_bringup_cmd = IncludeLaunchDescription(
        PythonLaunchDescriptionSource(
            os.path.join(get_package_share_directory('nav2_bringup'), 'launch'
                , 'bringup_launch.py')),
        launch_arguments = {
            'use_sim_time': 'False',
            'autostart': 'True',
            'use_composition': 'False',
            'map': map_file,
            'params_file': nav2_params_file}.items()
    )
    ld.add_action(nav2_bringup_cmd)

    robot_state_publisher_cmd = Node(
```

```python
            package = 'robot_state_publisher',
            executable = 'robot_state_publisher',
            name = 'tb3_state_publisher',
            output = 'screen',
            parameters = [{
                'use_sim_time': False,
                'robot_description': robot_description,
                'publish_frequency': 50.0
                }]
        )
    ld.add_action(robot_state_publisher_cmd)

    hlds_laser_publisher_cmd = Node(
            package = 'hls_lfcd_lds_driver',
            executable = 'hlds_laser_publisher',
            name = 'tb3_hlds_laser_publisher',
            output = 'screen',
            parameters = [{
                'port': '/dev/ttyUSB0',
                'frame_id': 'base_scan'
                }]
        )
    ld.add_action(hlds_laser_publisher_cmd)

    tb3_node_cmd = Node(
            package = 'turtlebot3_node',
            executable = 'turtlebot3_ros',
            output = 'screen',
            parameters=[tb3_param_file],
            arguments=['-i', '/dev/ttyACM0']
        )
    ld.add_action(tb3_node_cmd)

    tb3_move_to_goal_cmd = Node(
            package = 'chapter10_2_tb3',
            executable = 'nav2_move_to_goal',
            name = 'tb3_move_node',
            parameters=[{
                'ini_x': '0.0',
                'ini_y': '0.0'
                }]
        )
```

```
92        ld.add_action(tb3_move_to_goal_cmd)
93
94        return ld
```

上述 launch 文件的代码与第10.1.3节的代码类似,下面对其中不同的部分进行解释:
- 第 2~7 行加载了运行此 launch 文件需要用到的包;
- 第 25~28 行定义了地图参数文件,用于第 44 行加载 Nav 2 包时需要用到的地图参数,这里我们把相关的地图文件存放在当前的功能包文件夹里;
- 第 30~33 行定义了 Nav 2 需要用到的机器人参数文件,其中的参数主要是用于设置 Nav 2 导航功能,在第 45 行加载了此参数文件;
- 第 37~47 行通过加载 Nav 2 的 bringup_launch.py 节点文件实现导航功能;
- 第 83~92 行加载前面编写的利用 Nav 2 控制机器人定点移动的节点文件.

重新打开 chapter10_2_tb3 目录下的 setup.py 文件,将 data_files 域修改为如下形式:

```
1  data_files=[
2      ...
3      (os.path.join('share', package_name, 'launch'), glob(os.path.join('launch
       ', '*.launch.py'))),
4      (os.path.join('share', package_name, 'map'), glob(os.path.join('map', '
       *.*'))),
5  ],
```

并在文件头部添加如下代码:

```
1  import os
2  from glob import glob
```

由于 Nav 2 需要启动地图服务,需要相应的地图文件. 在 chapter10_2_tb3 目录下新建一个 map 目录,用于存放地图文件,运行如下命令:

```
$ cd ~/book_tb3/src/chapter10_2_tb3
$ mkdir map
```

并将地图文件(blank_map.pgm 和 blank_map.yaml)拷贝到 map 目录下. 注意这里使用了一张空的地图文件,即假设机器人是在一个空的环境中实现移动和编队任务. 如果读者需要实现在有障碍物环境中的机器人移动和编队任务,那么需要使用 SLAM (Simultaneous Localization and Mapping,同时定位与建图)技术,感兴趣的读者可以阅读相关资料.

接下来,重新构建 chapter10_2_tb3 功能包,运行如下命令:

```
$ cd ~/book_tb3
$ colcon build --packages-select chapter10_2_tb3
```

打开终端，输入如下命令进行测试：

```
$ ros2 launch chapter10_2_tb3 nav2_move_to_goal_demo.launch.py
```

从图10.3给出的 TurtleBot 3 机器人的实物测试结果可以看到机器人从$(0,0)$点出发，首先移动到$(0,1)$点，再到$(1,1)$点，最后移动到$(0,2)$点.

图 10.3　在实际环境中使用 Nav 2 实现 TurtleBot 3 机器人的定点移动

10.3　利用 Nav 2 实现多机器人队形切换

本节将基于第8章的仿真实验，利用 Nav 2 的导航功能，实现三个 TurtleBot 3 机器人的队形切换任务：从直线队形切换到三角形队形，然后再切换回到直线队形. 三个机器人形成一个连通的通信拓扑结构，如图10.4所示，图中的虚线表示机器人之间的通信关系. 从图中我们可以看出，robot1 的邻居是 robot3，robot2 的邻居也是 robot3，robot3 的邻居是 robot1 和 robot2.

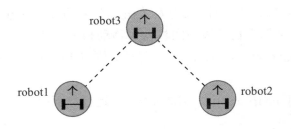

图 10.4　三个机器人形成连通的通信拓扑结构

我们将编写代码实现上述队形切换任务, 算法流程如图10.5所示. 这里我们采用域桥(domain bridge)[①] 的方式实现三个机器人在同一个域(domain0)内进行分布式优化计算, 分别在三个不同的域(domain1, domain2 和 domain3)内订阅目标位置信息并使用 Nav 2 实现定点移动. 每个机器人有两个节点程序文件, 一个主要用于实现分布式优化计算, 另一个主要用于实现 Nav 2 定点导航.

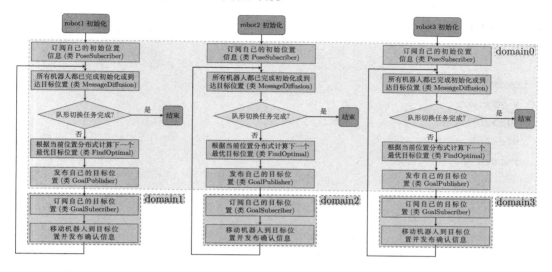

图 10.5　实现队形切换的分布式算法流程图

10.3.1　第一个机器人节点文件

在这一节中, 将具体编写实现上述设计框架的程序文件. 同样地, 使用 VS Code 软件远程登录第一个 TurtleBot 3 的 Ubuntu 系统. 新建一个终端窗口, 执行如下命令创建一个名称为 chapter10_3_tb3 的功能包:

```
$ cd ~/book_tb3/src
$ ros2 pkg create --build-type ament_python chapter10_3_tb3
```

然后构建这个包:

```
$ cd ~/book_tb3
$ colcon build --packages-select chapter10_3_tb3
$ . install/setup.bash
```

下面将基于第8章的仿真程序编写相应的实物实现代码. 在 chapter10_3_tb3 功能包中创建实现第一个机器人 robot1 进行分布式优化计算的 Python 节点文件:

```
$ cd ~/book_tb3/src/chapter10_3_tb3/chapter10_3_tb3
$ sudo nano robot_dis_find_opt.py
```

[①] 有关 domain bridge 的介绍可以参考官方文档(https://github.com/ros2/domain_bridge).

在 robot_dis_find_opt.py 文件中添加如下代码：

```python
#!/usr/bin/env python3
import rclpy
from rclpy.node import Node
from std_msgs.msg import Float32MultiArray, Int16, Int32MultiArray
import numpy as np
import time

def proj(x, lb, ub):
    y = x.copy()
    y[x>ub] = ub
    y[x<lb] = lb
    return y

class PoseSubscriber(Node):
    def __init__(self):
        super().__init__('pose_subscriber')
        self.declare_parameter('tb3name', '')
        self.tb3name = self.get_parameter('tb3name').get_parameter_value().\
            string_value
        self.pose_receive = None
        self.arrival_index = None

        self.sub_pose = self.create_subscription(
            Float32MultiArray,
            f'/{self.tb3name}/pose',
            self.pose_call_back,
            1)

        self.sub_confirm = self.create_subscription(
            Int16,
            f'/{self.tb3name}/arrival_index',
            self.arrival_confirm,
            1)

    def pose_call_back(self, msg):
        pose_array = msg.data
        self.pose_receive = pose_array.tolist()

    def arrival_confirm(self, msg):
        self.arrival_index = msg.data
```

```python
40
41          def pose_result(self):
42              return self.tb3name, self.pose_receive
43
44          def arrival_result(self):
45              return self.tb3name, self.arrival_index
46
47      class FindOptimal(Node):
48          def __init__(self, pose, Delta12, S):
49              super().__init__('find_optimal')
50              self.declare_parameter('tb3name', '')
51              self.declare_parameter('neighbor_name', '')
52              self.declare_parameter('sigma', '')
53              self.declare_parameter('upper_bound', '')
54              self.tb3name = self.get_parameter('tb3name').get_parameter_value().\
                    string_value
55              neighbor_name = self.get_parameter('neighbor_name').\
                    get_parameter_value().string_value
56              sigma = self.get_parameter('sigma').get_parameter_value().string_value
57              ub = self.get_parameter('upper_bound').get_parameter_value().\
                    string_value
58              neighbor_list = neighbor_name.split()
59              self.sigma = float(sigma)
60              self.ub = float(ub)
61
62              self.xi1 = np.array(pose).reshape(-1, 1) # 机器人1的初始位置常量\xi_1
63              self.b3 = np.array(Delta12).reshape(-1, 1) # 机器人1和机器人2的相对位置
                    常量b_3
64
65              self.p1 = np.zeros((2, 1)) # np.array(pose[1]).reshape(-1, 1) # 读取机
                    器人1 的位姿数据初始化优化变量p_1
66              self.p12 = np.zeros((2, 1)) # np.array(pose[2]).reshape(-1, 1) # 读取
                    机器人2 的位姿数据初始化优化变量p_12
67              self.y1 = np.zeros((2, 1)) # 辅助变量y_1初始值
68              self.z1 = np.zeros((2, 1)) # 辅助变量z_1初始值
69              self.w12 = np.zeros((2, 1)) # 辅助变量w_12初始值
70
71              self.p1_neighbor = 2 * np.random.rand(len(neighbor_list), 2) - 1 # 初
                    始化机器人邻居对p_1的估计
72              self.p2_neighbor = 2 * np.random.rand(len(neighbor_list), 2) - 1 # 初
                    始化机器人邻居对p_2的估计
73              for index, name in enumerate(neighbor_list):
```

```python
74                self.p1p2_monitor(index, name) # 初始化函数p1p2_monitor
75
76            self.pub_p1p2 = self.create_publisher( # 定义发布者用于发布机器人对p_1
                    和p_2 的估计值
77                Float32MultiArray,
78                f'/{self.tb3name}/p1p2_est',
79                1)
80
81            self.timer = self.create_timer(0.05, self.p1p2_pub) # 优化算法求解p_1
                    和p_2 的估计值并发布
82
83        def p1p2_monitor(self, index, name): # 订阅邻居对p1和p2的估计值
84            sub_p1p2 = self.create_subscription(
85                Float32MultiArray,
86                f'/{name}/p1p2_est',
87                lambda msg: self.p1p2_call_back(msg, index),
88                1)
89
90        def p1p2_call_back(self, msg, index):
91            self.p1_neighbor[index] = [msg.data[0], msg.data[1]]
92            self.p2_neighbor[index] = [msg.data[2], msg.data[3]]
93
94        def p1p2_pub(self):
95            p1_neighbor = np.array(self.p1_neighbor)
96            p2_neighbor = np.array(self.p2_neighbor)
97            p_1j = np.zeros((2,1))
98            p_2j = np.zeros((2,1))
99            for index in range(len(p1_neighbor)):
100               p_1j += self.p1-p1_neighbor[index].reshape(-1,1) # 自己与邻居对p_1
                        估计值差的累加
101               p_2j += self.p12-p2_neighbor[index].reshape(-1,1) # 自己与邻居对p_2
                        估计值差的累加
102
103           # 机器人1算法的迭代公式
104           p1 = proj(self.p1-self.sigma*(self.p1-self.xi1+self.y1+self.p1-self.
                    p12-self.b3+self.z1+p_1j), 0.0, self.ub)
105           p12 = proj(self.p12-self.sigma*(-self.y1-self.p1+self.p12+self.b3+self
                    .w12+p_2j), 0.0, self.ub)
106           y1 = self.y1+self.p1-self.p12-self.b3
107           z1 = self.z1+p_1j
108           w12 = self.w12+p_2j
109
```

```python
            self.p1 = p1
            self.p12 = p12
            self.y1 = y1
            self.z1 = z1
            self.w12 = w12
            current_p1p2 = np.array([self.p1, self.p12]).flatten()
            pub_p1p2_data = Float32MultiArray(data=current_p1p2)

            connections = self.pub_p1p2.get_subscription_count()
            if connections == len(p1_neighbor): # 如果所有的邻居连接成功，则发布当
                前的优化变量值
                self.pub_p1p2.publish(pub_p1p2_data)

    def opt_result(self):
        return self.tb3name, self.p1.flatten()

class GoalPublisher(Node):
    def __init__(self, goal, goal_index):
        super().__init__('goal_publisher')
        self.declare_parameter('tb3name', '')
        self.tb3name = self.get_parameter('tb3name').get_parameter_value().
            string_value
        self.arrival_index = None
        self.goal = np.append(goal, goal_index)

        self.pub_goal = self.create_publisher(
            Float32MultiArray,
            f'/{self.tb3name}/goal',
            1)

        self.sub_confirm = self.create_subscription(
            Int16,
            f'/{self.tb3name}/arrival_index',
            self.arrival_confirm,
            1)

        self.timer1 = self.create_timer(0.01, self.goal_pub) # 发布机器人目标
            位置坐标

    def goal_pub(self):
        pub_goal_data = Float32MultiArray(data=self.goal)
        self.pub_goal.publish(pub_goal_data)
```

```python
    def arrival_confirm(self, msg):
        self.arrival_index = msg.data

    def arrival_result(self):
        return self.tb3name, self.arrival_index

class MessageDiffusion(Node):
    def __init__(self):
        super().__init__('message_diffusion')
        self.declare_parameter('tb3name', '')
        self.declare_parameter('neighbor_name', '')
        tb3name = self.get_parameter('tb3name').get_parameter_value().\
            string_value
        neighbor_name = self.get_parameter('neighbor_name').\
            get_parameter_value().string_value
        self.neighbor_list = neighbor_name.split()
        robot_id = int(tb3name[-1])

        self.msg_pub = [robot_id] # 记录发布的信息
        self.pub_arriver = self.create_publisher( # 定义发布者用于发布到达目标
            位置的确认信息
            Int32MultiArray,
            f'/{tb3name}/arriver_list',
            1)

        for name in self.neighbor_list:
            self.arriver_monitor(name)

        arriver_pub_timer = self.create_timer(0.1, self.arriver_pub) # 到达目
            标位置后发布确认信息

    def arriver_monitor(self, name): # 订阅邻居的确认信息
        sub_arriver = self.create_subscription(
            Int32MultiArray,
            f'/{name}/arriver_list',
            self.arriver_call_back,
            1)

    def arriver_call_back(self, msg):
        self.msg_pub = list(set(self.msg_pub+list(msg.data)))

```

```python
    def arriver_pub(self):
        pub_arriver_data = Int32MultiArray(data=self.msg_pub)
        connections = self.pub_arriver.get_subscription_count()
        if connections == len(self.neighbor_list):
            self.pub_arriver.publish(pub_arriver_data)

    def arriver_result(self, robot_number):
        all_arrival = False
        if len(self.msg_pub) == robot_number:  # 如果所有机器人都到达目标位置，
                                               # 则返回逻辑值True
            all_arrival = True
        return all_arrival

def main():
    rclpy.init()
    node = PoseSubscriber()
    node.declare_parameter('robot_number', '')
    node.declare_parameter('max_time', '')
    robot_number = node.get_parameter('robot_number').get_parameter_value().\
        string_value
    max_time = node.get_parameter('max_time').get_parameter_value().\
        string_value
    robot_number = int(robot_number)
    max_time = float(max_time)
    while rclpy.ok():
        try:
            rclpy.spin_once(node)
            tb3name, pose_now = node.pose_result()
            tb3name, arrival_index = node.arrival_result()
        except KeyboardInterrupt:
            pass
        if pose_now is not None and arrival_index is not None:
            node.get_logger().info('{0}初始位置已经收到，将要计算目标位置.'.
                format(tb3name))
            node.get_logger().info('机器人到达次数索引：{0}'.format(
                arrival_index))
            break
    node.destroy_node()
    time.sleep(1.0)

    node = MessageDiffusion()
    delay_counter = 0  # 延迟计数器
```

```
224        while rclpy.ok():
225            try:
226                rclpy.spin_once(node)
227                all_arrival = node.arriver_result(robot_number)
228            except KeyboardInterrupt:
229                pass
230            if all_arrival is True:
231                delay_counter += 1
232                if delay_counter == 20:
233                    node.get_logger().info('所有机器人都已到达目标位置.')
234                    break
235        node.destroy_node()
236
237        # 队形设置
238        S = np.zeros((2, robot_number, 2))
239        S[0] = np.array([[0, 0], [1, 0], [0, 1]])
240        S[1] = np.array([[0, 0], [1, 0], [2, 0]])
241        Delta12 = [[-0.8, -0.8], [-0.5, 0.5]]
242
243        for i in range(len(S)):
244            node = FindOptimal(pose_now, Delta12[i], S[i])
245            time_start, _ = node.get_clock().now().seconds_nanoseconds()
246            while rclpy.ok():
247                try:
248                    rclpy.spin_once(node)
249                    tb3name, goal = node.opt_result()
250                except KeyboardInterrupt:
251                    pass
252                time_now, _ = node.get_clock().now().seconds_nanoseconds()
253                if time_now - time_start > max_time:
254                    node.get_logger().info('{0}得到的最优解: {1}'.format(tb3name,
                        goal))
255                    break
256            node.destroy_node()
257
258            arrival_index_previous = arrival_index
259            node = GoalPublisher(goal, i)
260            while rclpy.ok():
261                try:
262                    rclpy.spin_once(node)
263                    tb3name, arrival_index = node.arrival_result()
264                except KeyboardInterrupt:
```

```
265                    pass
266                if arrival_index is not None and arrival_index >
                       arrival_index_previous:
267                    node.get_logger().info('{0}到达目标位置，将计算下一个目标位置.'.
                           format(tb3name))
268                    node.get_logger().info('机器人到达次数索引：{0}'.format(
                           arrival_index))
269                    break
270        node.destroy_node()
271
272        node = MessageDiffusion()
273        delay_counter = 0 # 延迟计数器
274        while rclpy.ok():
275            try:
276                rclpy.spin_once(node)
277                all_arrival = node.arriver_result(robot_number)
278            except KeyboardInterrupt:
279                pass
280            if all_arrival is True:
281                delay_counter += 1
282                if delay_counter == 20:
283                    node.get_logger().info('所有机器人都已到达目标位置.')
284                    break
285        node.destroy_node()
286        pose_now = goal[0:2]
287    rclpy.shutdown()
288
289 if __name__ == '__main__':
290     main()
```

这里的代码与8.2节相似，下面对其中不同的地方进行解释：

- 第 2~6 行加载了需要用到的包；
- 第 14~45 行定义了类 PoseSubscriber()，用于跨域订阅机器人的初始位置信息和机器人到达目标位置的确认信息；
- 第 47~123 行定义了类 FindOptimal()，用于分布式计算最优队形所对应的目标位置；
- 第 125~154 行定义了类 GoalPublisher()，用于跨域发布机器人的目标位置信息和订阅机器人到达目标位置的确认信息；
- 第 156~197 行定义了类 MessageDiffusion()，用于确认所有机器人是否到达目标位置；
- 第 199~287 行定义了主函数，主要包括以下内容：

- 第 201~219 行实例化类 PoseSubscriber(),跨域订阅机器人的初始位置信息,并确认机器人已经完成初始化;
- 第 222~235 行实例化类 MessageDiffusion(),用于确认所有机器人都已完成初始化,以同步启动分布式优化计算节点;
- 第 238~241 行是机器人队形的设置参数;
- 第 244~256 行实例化类 FindOptimal(),根据机器人当前的位置信息分布式计算最优队形,并将机器人的目标位置存储在变量 goal 中;
- 第 258~270 行实例化类 GoalPublisher(),用于跨域发布之前计算得到的最优目标位置,并同时订阅机器人到达上一个目标位置的确认信息;
- 第 272~286 行实例化类 MessageDiffusion(),用于确认所有机器人都已到达目标位置,以同步启动下一次的分布式优化计算节点.

此外,还要创建一个实现 Nav 2 定点导航功能的 Python 节点文件:

```
$ cd ~/book_tb3/src/chapter10_3_tb3/chapter10_3_tb3
$ sudo nano robot_move_to_goal_nav2.py
```

在 robot_move_to_goal_nav2.py 文件中添加如下代码:

```python
#!/usr/bin/env python3
import rclpy
from rclpy.node import Node
from geometry_msgs.msg import PoseStamped
from nav2_simple_commander.robot_navigator import BasicNavigator, TaskResult
from std_msgs.msg import Float32MultiArray, Int16
import time

class GoalSubscriber(Node):
    def __init__(self, pose, arrival_index):
        super().__init__('goal_subscriber')
        self.declare_parameter('tb3name', '')
        self.tb3name = self.get_parameter('tb3name').get_parameter_value().
            string_value
        self.pose_send = pose
        self.goal_receive = None
        self.arrival_index = arrival_index
        self.get_logger().info('机器人到达次数索引: {0}'.format(self.
            arrival_index))

        self.pub_pose = self.create_publisher(
            Float32MultiArray,
            f'/{self.tb3name}/pose',
            1)
```

```python
        self.sub_goal = self.create_subscription(
            Float32MultiArray,
            f'/{self.tb3name}/goal',
            self.goal_call_back,
            1)

        self.pub_arrival = self.create_publisher(
            Int16,
            f'/{self.tb3name}/arrival_index',
            1)

        self.timer1 = self.create_timer(0.01, self.pose_pub) # 发布机器人当前
            位置坐标
        self.timer2 = self.create_timer(0.01, self.arrival_pub) # 发布机器人到
            达目标位置的时间

    def pose_pub(self):
        pub_pose_data = Float32MultiArray(data=self.pose_send)
        self.pub_pose.publish(pub_pose_data)

    def goal_call_back(self, msg):
        goal_array = msg.data
        self.goal_receive = goal_array.tolist()

    def arrival_pub(self):
        pub_arrival_data = Int16(data=self.arrival_index)
        self.pub_arrival.publish(pub_arrival_data)

    def goal_result(self):
        return self.tb3name, self.goal_receive

def main():
    rclpy.init()
    navigator = BasicNavigator()
    navigator.declare_parameter('ini_x', '')
    navigator.declare_parameter('ini_y', '')
    ini_x = navigator.get_parameter('ini_x').get_parameter_value().\
        string_value
    ini_y = navigator.get_parameter('ini_y').get_parameter_value().\
        string_value
    ini_x = float(ini_x)
```

```python
ini_y = float(ini_y)

# 设置机器人的初始位姿
initial_pose = PoseStamped()
initial_pose.header.frame_id = 'map'
initial_pose.header.stamp = navigator.get_clock().now().to_msg()
initial_pose.pose.position.x = ini_x
initial_pose.pose.position.y = ini_y
navigator.setInitialPose(initial_pose)

navigator.waitUntilNav2Active() # 等待激活Nav2

pose_now = [ini_x, ini_y]
# 实例化GoalSubscriber()对象获取机器人初始目标位置
node = GoalSubscriber(pose_now, 0)
while rclpy.ok():
    try:
        rclpy.spin_once(node)
        tb3name, goal = node.goal_result()
    except KeyboardInterrupt:
        pass
    if goal is not None:
        break
node.destroy_node()

for i in range(2):
    goal_x = goal[0]
    goal_y = goal[1]
    # 移动机器人到指定目标位置
    goal_pose = PoseStamped()
    goal_pose.header.frame_id = 'map'
    goal_pose.header.stamp = navigator.get_clock().now().to_msg()
    goal_pose.pose.position.x = goal_x
    goal_pose.pose.position.y = goal_y
    time.sleep(1.0)
    navigator.goToPose(goal_pose)
    count = 0
    while not navigator.isTaskComplete():
        count += 1
        if count % 10 == 0:
            navigator.get_logger().info('{0}正在前往目标位置: {1}'.format(
                tb3name, goal))
```

```
102
103                # 根据返回结果输出相应信息或者订阅新的目标位置
104                result = navigator.getResult()
105                if result == TaskResult.SUCCEEDED:
106                    navigator.get_logger().info('{0}成功到达目标位置!'.format(tb3name))
107                    # 实例化GoalSubscriber()对象获取机器人下一个目标位置
108                    pose_now = goal[0:2]
109                    goal_previous = goal
110                    node = GoalSubscriber(pose_now, i+1)
111                    while rclpy.ok():
112                        try:
113                            rclpy.spin_once(node)
114                            tb3name, goal = node.goal_result()
115                        except KeyboardInterrupt:
116                            pass
117                        if goal is not None and goal[2] > goal_previous[2]: # 如果收到
                                新的目标位置
118                            break
119                    node.destroy_node()
120                elif result == TaskResult.CANCELED:
121                    navigator.get_logger().info('{0}目标位置被取消!'.format(tb3name))
122                elif result == TaskResult.FAILED:
123                    navigator.get_logger().info('{0}未能到达目标位置!'.format(tb3name))
124                else:
125                    navigator.get_logger().info('{0}目标有一个无效的返回状态!'.format(
                        tb3name))
126        navigator.destroy_node()
127        rclpy.shutdown()
128
129   if __name__ == '__main__':
130       main()
```

这里的代码与第10.2.1节相似，下面对不同的地方进行解释：

- 第2~7行加载了运行此文件需要用到的包，和之前的代码用到的包基本相同；
- 第9~51行定义了类GoalSubscriber()，同时启动了多个订阅者和发布者，用于跨域订阅机器人的目标位置信息，并同时发布机器人当前的位置信息和机器人到达目标位置时的索引信息，其中索引信息是指：当机器人完成初始化时，其索引值为0；当机器人第一次到达目标位置时，其索引值为1；当机器人第二次到达目标位置时，其索引值为2，以此类推；
- 第53~127行定义了主函数，主要包括以下内容：
  - 第55~71行实例化类BasicNavigator()，实现机器人的初始化和等待Nav 2

- 导航包完成激活;
- 第 73~84 行实例化类 GoalSubscriber(),机器人发布初始位置信息的同时订阅第一个目标位置;
- 第 86~125 行启动一个循环,当机器人接收到新的目标位置时,即通过 Nav 2 导航功能实现定点移动,其中第 104~125 行根据返回结果输出相应信息,当机器人成功到达目标位置时,重新实例化类 GoalSubscriber() 订阅新的目标位置.

打开 chapter10_3_tb3 目录下的 setup.py 文件,将 entry_points 域修改为如下形式:

```
entry_points={
    'console_scripts': [
        'robot_dis_find_opt = chapter10_3_tb3.robot_dis_find_opt:main',
        'robot_move_to_goal_nav2 = chapter10_3_tb3.robot_move_to_goal_nav2:
            main',
    ],
},
```

10.3.2 第一个机器人 launch 文件

由于需要在两个域分别运行分布式优化计算和控制机器人移动,因此需要编写两个相应的 launch 文件. 首先编写一个 launch 文件加载分布式优化计算的节点文件. 接下来,在 chapter10_3_tb3 目录下新建一个 launch 目录,用于存放 launch 文件. 运行如下命令:

```
$ cd ~/book_tb3/src/chapter10_3_tb3
$ mkdir launch
```

在 launch 目录下新建一个名称为 tb3_find_optimal.launch.py 的文件:

```
$ cd ~/book_tb3/src/chapter10_3_tb3/launch
$ sudo nano tb3_find_optimal.launch.py
```

并添加如下代码:

```
#!/usr/bin/env python3
import os
from launch import LaunchDescription
from launch_ros.actions import Node

def generate_launch_description():
    robot = {'name': 'robot1', 'neighbor': 'robot3'}
    robot_number = '3' # 机器人数量
```

```
9       max_time = '30' # 分布式优化算法最大运行时间
10      sigma = '0.05' # 优化算法步长\sigma
11      upper_bound = '2.0' # 编队移动范围的上界
12
13      ld = LaunchDescription()
14
15      robot_distributed_opt_cmd = Node(
16          package = 'chapter10_3_tb3',
17          executable = 'robot_dis_find_opt',
18          namespace = robot.get('name'),
19          name = 'distributed_node',
20          output = 'screen',
21          parameters = [{
22              'tb3name': robot.get('name'),
23              'neighbor_name': robot.get('neighbor'),
24              'robot_number': robot_number,
25              'max_time': max_time,
26              'sigma': sigma,
27              'upper_bound': upper_bound
28              }],
29          )
30      ld.add_action(robot_distributed_opt_cmd)
31
32      return ld
```

这里的代码与第8.2.2节相似，不同的地方是这里加载了robot_dis_find_opt可执行节点，用于实现分布式优化计算.

接下来，新建另一个launch文件用于启动Nav 2包并实现机器人定点移动. 在launch目录下新建一个名称为nav2_move_to_goal.launch.py的文件：

```
$ cd ~/book_tb3/src/chapter10_3_tb3/launch
$ sudo nano nav2_move_to_goal.launch.py
```

并添加如下代码：

```
1  #!/usr/bin/env python3
2  import os
3  from ament_index_python.packages import get_package_share_directory
4  from launch import LaunchDescription
5  from launch.actions import IncludeLaunchDescription
6  from launch.launch_description_sources import PythonLaunchDescriptionSource
7  from launch_ros.actions import Node
8
```

```python
 9  def generate_launch_description():
10      TURTLEBOT3_MODEL = os.environ['TURTLEBOT3_MODEL']
11
12      tb3_param_file = os.path.join(
13          get_package_share_directory('turtlebot3_bringup'),
14          'param',
15          TURTLEBOT3_MODEL + '.yaml')
16
17      urdf_file_name = 'turtlebot3_' + TURTLEBOT3_MODEL + '.urdf'
18      urdf_path = os.path.join(
19          get_package_share_directory('turtlebot3_description'),
20          'urdf',
21          urdf_file_name)
22      with open(urdf_path, 'r') as infp:
23          robot_description = infp.read()
24
25      map_file = os.path.join(
26          get_package_share_directory('chapter10_2_tb3'),
27          'map',
28          'blank_map.yaml')
29
30      nav2_params_file = os.path.join(
31          get_package_share_directory('turtlebot3_navigation2'),
32          'param',
33          'burger.yaml')
34
35      robot = {'name': 'robot1', 'x_pose': '0.0', 'y_pose': '0.0'}
36
37      ld = LaunchDescription()
38
39      nav2_bringup_cmd = IncludeLaunchDescription(
40          PythonLaunchDescriptionSource(
41              os.path.join(get_package_share_directory('nav2_bringup'), 'launch'
                  , 'bringup_launch.py')),
42          launch_arguments = {
43              'use_sim_time': 'False',
44              'autostart': 'True',
45              'use_composition': 'False',
46              'map': map_file,
47              'params_file': nav2_params_file}.items()
48          )
49      ld.add_action(nav2_bringup_cmd)
```

```
50
51      robot_state_publisher_cmd = Node(
52          package = 'robot_state_publisher',
53          executable = 'robot_state_publisher',
54          name = 'tb3_state_publisher',
55          output = 'screen',
56          parameters = [{
57              'use_sim_time': False,
58              'robot_description': robot_description,
59              'publish_frequency': 50.0
60              }]
61          )
62      ld.add_action(robot_state_publisher_cmd)
63
64      hlds_laser_publisher_cmd = Node(
65          package = 'hls_lfcd_lds_driver',
66          executable = 'hlds_laser_publisher',
67          name = 'tb3_hlds_laser_publisher',
68          output = 'screen',
69          parameters = [{
70              'port': '/dev/ttyUSB0',
71              'frame_id': 'base_scan'
72              }]
73          )
74      ld.add_action(hlds_laser_publisher_cmd)
75
76      tb3_node_cmd = Node(
77          package = 'turtlebot3_node',
78          executable = 'turtlebot3_ros',
79          output = 'screen',
80          parameters = [tb3_param_file],
81          arguments = ['-i', '/dev/ttyACM0']
82          )
83      ld.add_action(tb3_node_cmd)
84
85      tb3_move_to_goal_cmd = Node(
86          package = 'chapter10_3_tb3',
87          executable = 'robot_move_to_goal_nav2',
88          name = 'tb3_move_node',
89          output = 'screen',
90          parameters = [{
91              'tb3name': robot.get('name'),
```

```
92              'ini_x': robot.get('x_pose'),
93              'ini_y': robot.get('y_pose')
94          }],
95      )
96      ld.add_action(tb3_move_to_goal_cmd)
97
98      return ld
```

这里的代码与第10.2.2节基本相同,下面不再进行详细解释.

下面重新打开 chapter10_3_tb3 目录下的 setup.py 文件,将 data_files 域修改为如下形式:

```
1  data_files=[
2      ...
3      (os.path.join('share', package_name, 'launch'), glob(os.path.join('launch
          ', '*.launch.py'))),
4  ],
```

并在文件头部添加如下代码:

```
1  import os
2  from glob import glob
```

10.3.3 第一个机器人配置文件

接下来,为了能够使用域桥功能,在 chapter10_3_tb3 目录下新建一个 config 目录,用于存放配置文件. 运行如下命令:

```
$ cd ~/book_tb3/src/chapter10_3_tb3
$ mkdir config
```

在 config 目录下新建一个名称为 domain_bridge_config.yaml 的文件:

```
$ cd ~/book_tb3/src/chapter10_3_tb3/config
$ sudo nano domain_bridge_config.yaml
```

并添加如下代码:

```
1  name: domain_bridge
2  topics:
3    /robot1/pose:
4      type: std_msgs/msg/Float32MultiArray
5      from_domain: 1
6      to_domain: 0
7    /robot1/goal:
```

```
 8        type: std_msgs/msg/Float32MultiArray
 9        from_domain: 0
10        to_domain: 1
11      /robot1/arrival_index:
12        type: std_msgs/msg/Int16
13        from_domain: 1
14        to_domain: 0
```

重新打开 chapter10_3_tb3 目录下的 setup.py 文件，将 data_files 域修改为如下形式：

```
1  data_files=[
2      ...
3      (os.path.join('share', package_name, 'launch'), glob(os.path.join('launch
       ', '*.launch.py'))),
4      (os.path.join('share', package_name, 'config'), glob(os.path.join('config
       ', '*.yaml'))),
5  ],
```

接下来，重新构建 chapter10_3_tb3 功能包，运行如下命令：

```
$ cd ~/book_tb3
$ colcon build --packages-select chapter10_3_tb3
```

为了方便代码的批量运行，在 chapter10_3_tb3 目录下新建一个名称为 auto_boot.sh 的 shell 脚本文件：

```
$ cd ~/book_tb3/src/chapter10_3_tb3
$ sudo nano auto_boot.sh
```

并添加如下代码：

```
1  #!/bin/bash
2  ros2 run domain_bridge domain_bridge ~/book_tb3/src/chapter10_3_tb3/config/
       domain_bridge_config.yaml &
3  ROS_DOMAIN_ID=0 ros2 launch chapter10_3_tb3 tb3_find_optimal.launch.py &
4  ROS_DOMAIN_ID=1 ros2 launch chapter10_3_tb3 nav2_move_to_goal.launch.py
```

至此，第一个机器人 robot1 的所有代码和配置文件都完成了．对于另外两个机器人 robot2 和 robot3，其代码和配置与 robot1 类似，不同的地方是用公式(3.26)和(3.27)替换 robot1 代码的类 FindOptimal() 的相应部分，由于其内容与 robot1 相似，这里就不再详细阐述．

10.3.4 实物测试

将三个机器人的初始位置依次设置为 $(0,0)$, $(0,-0.3)$ 和 $(0,0.3)$, 即三个机器人从直线队形出发, 首先变换成三角形队形, 进而再变回直线队形. 可以通过远程连接的方式分别登录三个 TurtleBot 3 机器人的 Ubuntu 系统, 并在终端输入如下命令启动前面编写的脚本文件:

```
$ ~/book_tb3/src/chapter10_3_tb3/auto_boot.sh
```

此外, 也可以使用另外一种更简洁的方式同时启动三个 TurtleBot 3 机器人的脚本文件. 打开本地电脑的 ubuntu 系统, 运行如下命令新建一个 shell 脚本文件:

```
$ cd ~/book_ws/src
$ mkdir chapter10_3
$ sudo nano multi_tb3_remote_boot.sh
```

并在 multi_tb3_remote_boot.sh 文件中添加如下代码:

```
#!/bin/bash

ip_array=("192.168.0.63" "192.168.0.64" "192.168.0.65")
user=ubuntu
password=turtlebot

for ip in ${ip_array[*]}
do
    sshpass -p $password ssh -tt $user@$ip > /dev/null 2>&1 << remotessh
    ~/book_tb3/src/chapter10_3_tb3/auto_boot.sh &
    exit
remotessh
done
```

该文件主要用于远程登录 TurtleBot 3 的 Ubuntu 系统, 并依次运行前面编写的三个机器人的 shell 脚本文件. 注意读者在使用此脚本时, 需要将代码第 3 行的 ip_array 替换成自己的 ip 地址.

打开本地终端, 输入如下命令进行测试:

```
$ cd ~/book_ws/src/chapter10_3
$ ./multi_tb3_remote_boot.sh
```

从图10.6给出的在实际环境中的测试结果可以看到三个机器人从直线队形出发, 先后切换成三角形队形和直线队形.

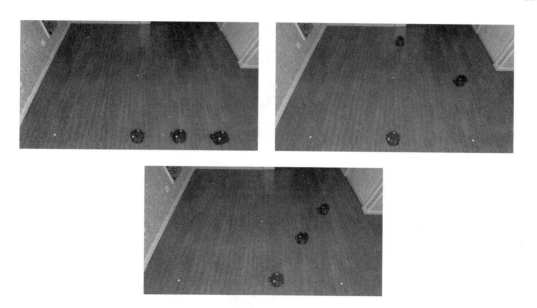

图 10.6　在真实环境中实现三个 TurtleBot 3 机器人的队形切换

练　习

1. 参考本章的代码，结合第7章的仿真实验，编写代码实现三个实物 TurtleBot 3 机器人的队形切换.
2. (切换拓扑下的多机器人编队)考虑如图10.7所示的两个通信拓扑结构，假设机器人的通信网络在这两个拓扑结构下每隔 1 秒钟切换一次，试参考本章的代码，实现切换拓扑下的多机器人队形切换.

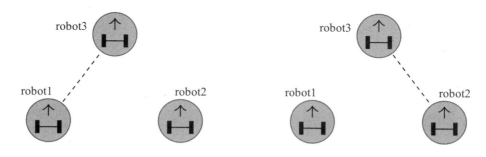

图 10.7　三个机器人形成的两个切换通信拓扑结构

附录 A 微分系统

本附录主要介绍微分系统的一些基本概念, 以及对微分系统的常用数值解法.

A.1 微分系统[①]

动力系统(dynamical system)是研究随时间演变的系统的一门分支学科, 又称动力学系统或动态系统. **微分动力系统**(differential dynamical system)是用微分方程描述的动力系统, 包括常微分方程和偏微分方程. 这里主要考虑常微分方程.

接下来, 通过几个实际的例子, 直观理解微分方程的概念.

例 A.1 平面上一曲线通过点 $(1, 2)$, 在该曲线上任意点处的切线斜率为 $2x$, 求该曲线的方程.

解 设所求曲线方程为 $y = y(x)$. 由微积分可知, 一元函数的导数的几何意义是函数所表示的曲线斜率, 因此有如下关系式:

$$\begin{cases} \dfrac{\mathrm{d}y}{\mathrm{d}x} = 2x \\ y(1) = 2 \end{cases}$$

上述的第一个等式即确定了一个微分方程, 第二个等式是该微分方程需要满足的条件. 对该微分方程两边同时积分, 很容易求得该微分方程的解为 $y = x^2 + 1$, 此即为满足条件的曲线方程.

上述的例A.1仅含有一个自变量, 通常称为一元常微分方程(简称常微分方程). 接下来看一个有多个自变量的例子.

例 A.2 (洛伦兹(Lorenz)系统) 将描述大气热对流的非线性微分方程组简化成描述

[①] 有关微分系统的详细介绍可以参考文献[17].

垂直速度和上下温差的展开系数 x, y, z 与时间 t 的关系:

$$\begin{cases} \dfrac{\mathrm{d}x}{\mathrm{d}t} = -a(x-y) \\ \dfrac{\mathrm{d}y}{\mathrm{d}t} = (b-z)x - y \\ \dfrac{\mathrm{d}z}{\mathrm{d}t} = xy - cz \end{cases}$$

其中 a, b, c 是常系数. 此微分系统通常称为洛伦兹系统, 对该系统的研究引起了对混沌研究的热潮.

洛伦兹系统具有非线性特征, 无法像例A.1那样求得其解析解. 然而, 可以通过数值方法得到其数值解. 如图A.1所示, 取 $a = 10, b = 28, c = 8/3$, 通过微分方程数值解法画出洛伦兹系统的三维演化轨迹.

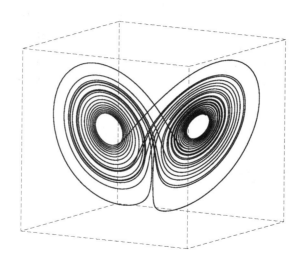

图 A.1　洛伦兹系统的三维演化轨迹

接下来给出微分方程和微分方程组的定义.

定义 A.1

含未知函数及其导数的方程叫作微分方程. 含有多个函数的微分方程构成微分方程组, 其一般形式为

$$\begin{cases} \dfrac{\mathrm{d}x_1}{\mathrm{d}t} = f_1(x_1, x_2, \cdots, x_n, t) \\ \dfrac{\mathrm{d}x_2}{\mathrm{d}t} = f_2(x_1, x_2, \cdots, x_n, t) \\ \cdots \\ \dfrac{\mathrm{d}x_n}{\mathrm{d}t} = f_n(x_1, x_2, \cdots, x_n, t) \end{cases}$$

若记 $\boldsymbol{x} = (x_1, x_2, \cdots, x_n)^{\mathrm{T}}$, $\boldsymbol{f} = (f_1, f_2, \cdots, f_n)^{\mathrm{T}}$, 则微分方程组可以写成如下的向量形式:

$$\frac{\mathrm{d}\boldsymbol{x}}{\mathrm{d}t} = \boldsymbol{f}(\boldsymbol{x}, t)$$

一般说, 大多数的微分方程(组)都无法得到其解析解, 因此需要用数值方法对微分方程进行求解.

A.2 微分系统数值方法[①]

这里, 介绍微分方程的两个常用的数值方法: 一个是欧拉(Euler)法, 另一个是龙格–库塔(Runge-Kutta)法.

对于如下的一般向量形式的微分系统:

$$\frac{\mathrm{d}\boldsymbol{x}}{\mathrm{d}t} = \boldsymbol{f}(\boldsymbol{x}, t) \tag{A.1}$$

其中 $x \in \mathbf{R}^n$, 且初始值满足 $x(t_0) = x_0$.

将向量 $x(t)$ 的导数近似表示为

$$\frac{\mathrm{d}\boldsymbol{x}(t)}{\mathrm{d}t} \approx \frac{\boldsymbol{x}(t+h) - \boldsymbol{x}(t)}{h}$$

即微分系统(A.1)可以近似表示为

$$\boldsymbol{x}(t+h) \approx \boldsymbol{x}(t) + h\boldsymbol{f}(\boldsymbol{x}, t)$$

其中 h 是步长. 记 t 是第 k 步迭代, $t+h$ 是第 $k+1$ 步迭代, 那么可以得到如下的欧拉迭代公式:

$$\boldsymbol{x}(k+1) = \boldsymbol{x}(k) + h\boldsymbol{f}(\boldsymbol{x}(k), k)$$

当步长 h 取值足够小时, 欧拉迭代公式可以用于近似计算微分系统(A.1)的状态向量 $\boldsymbol{x}(t)$.

当使用欧拉迭代公式计算微分系统时, 由于步长比较小, 往往会使得算法的收敛速度非常慢. 为了克服这个问题, 介绍另一个常用的微分系统数值方法, 即龙格–库塔法.

[①] 有关微分系统数值方法的详细介绍可以参考文献[14].

对于微分系统(A.1), 龙格–库塔算法的迭代公式如下所示:

$$\begin{cases} r_1 = f(x_k, t_k) \\ r_2 = f(x_k + \frac{h}{2}r_1, t_k + \frac{h}{2}) \\ r_3 = f(x_k + \frac{h}{2}r_2, t_k + \frac{h}{2}) \\ r_4 = f(x_k + hr_3, t_k + h) \end{cases}$$

$$x_{k+1} = \frac{h}{6}(r_1 + 2r_2 + 2r_3 + r_4)$$

其中 h 是步长. 相比于欧拉方法, 龙格–库塔方法的优势是使用了更多的差值点, 因此可以得到更高的计算精度和更快的收敛速度.

附录 B 李雅普诺夫稳定性理论

本附录主要介绍微分系统定性分析中的李雅普诺夫(Lyapunov)稳定性理论. 早在 1892 年, 俄国学者李雅普诺夫(Aleksandr Mikhailovich Lyapunov, 1857~1918)发表题为 "运动稳定性一般问题" 的著名文献, 建立了关于运动稳定性研究的一般理论. 百余年来, 李雅普诺夫理论得到极大发展, 在数学、力学、控制理论、机械工程等领域得到广泛应用.

B.1 稳定性定义

李雅普诺夫把分析一阶常微分方程组稳定性的所有方法归纳为两类:
- 第一类方法是将非线性系统在平衡态附近线性化, 然后通过讨论线性化系统的特征值(或极点)分布及稳定性来讨论原非线性系统的稳定性问题, 称为**间接法**, 亦称为**李雅普诺夫第一法**;
- 第二类方法不是通过解方程或求系统特征值来判别稳定性, 而是通过定义一个叫作**李雅普诺夫函数**的标量函数来分析判别稳定性, 称为**直接法**, 亦称为**李雅普诺夫第二法**.

这里主要介绍李雅普诺夫第二法.

对于如下的一般向量形式的微分系统:

$$\frac{\mathrm{d}\boldsymbol{x}}{\mathrm{d}t} = \boldsymbol{f}(\boldsymbol{x}, t) \tag{B.1}$$

其中 $\boldsymbol{x} \in \mathbf{R}^n$.

> **定义 B.1 (平衡点 (平衡态))**
>
> 微分系统(B.1)的平衡点(平衡态) 是指对所有的 t 满足
>
> $$\boldsymbol{f}(\boldsymbol{x}, t) = 0$$
>
> 的状态, 用 $\boldsymbol{x}_\mathrm{e}$ 来表示, 即 $\boldsymbol{f}(\boldsymbol{x}_\mathrm{e}, t) = 0$.

附录 B 李雅普诺夫稳定性理论

李雅普诺夫稳定性理论研究微分系统在平衡点附近(邻域)内的运动变化问题. 若平衡点附近某充分小邻域内所有状态的运动最后都趋于该平衡点, 则称该平衡点是**渐近稳定**的; 若发散则称该平衡点是**不稳定**的; 若能维持在平衡点附近某个邻域内运动变化则称该平衡点是**稳定**的. 下面给出上述稳定性定义的严格数学描述.

定义 B.2 (稳定 (stability))

若状态方程(B.1)所描述的微分系统满足, 对于任意的 $\varepsilon > 0$ 和任意初始时刻 t_0, 都对应存在一个实数 $\delta(\varepsilon, t_0) > 0$, 使得对于任意位于平衡点 $\boldsymbol{x}_e$ 的邻域 $N(\boldsymbol{x}_e, \delta)$ 内的初始状态 $\boldsymbol{x}_0$, 当从此初始状态 $\boldsymbol{x}_0$ 出发的状态方程的解 $\boldsymbol{x}$ 都位于邻域 $N(\boldsymbol{x}_e, \varepsilon)$ 内, 则称微分系统的平衡点 $\boldsymbol{x}_e$ 是李雅普诺夫意义下稳定的.

定义 B.3 (渐近稳定, asymptotic stability)

若状态方程(B.1)所描述的微分系统在初始时刻 t_0 的平衡点 $\boldsymbol{x}_e$ 是李雅普诺夫意义下稳定的, 且系统状态最终趋近于系统的平衡点 $\boldsymbol{x}_e$, 即 $\lim\limits_{t \to \infty} \boldsymbol{x}(t) = \boldsymbol{x}_e$, 则称微分系统的平衡点 $\boldsymbol{x}_e$ 是李雅普诺夫意义下渐近稳定的, 其中的满足极限行为称为收敛或者吸引.

如图B.1所示, 左图给出了稳定的示意图, 即微分系统的状态曲线从初始点 $\boldsymbol{x}_0$ 出发, 只要 $\boldsymbol{x}_0$ 在平衡点 $\boldsymbol{x}_e$ 的邻域 $N(\boldsymbol{x}_e, \delta)$ 内, 系统状态 $x(t)$ 就位于邻域 $N(\boldsymbol{x}_e, \varepsilon)$ 内; 右图给出了渐近稳定的示意图, 即平衡点 $\boldsymbol{x}_e$ 是稳定的, 并且满足当 $t \to \infty$ 时 $\boldsymbol{x}(t) \to \boldsymbol{x}_e$.

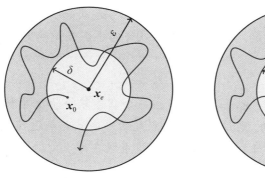
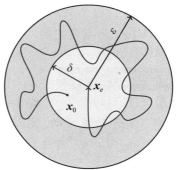

图 B.1 稳定与渐近稳定示意图

此外, 对于 n 维状态空间中的所有状态, 如果由这些初始状态出发的状态轨线都具有渐近稳定性, 那么平衡点 $\boldsymbol{x}_e$ 称为李雅普诺夫意义下**全局渐近稳定**的. 换句话说, 若状态方程(B.1)在任意初始状态下的解, 当 t 无限增长时都趋于平衡点, 则该平衡点为全局渐近稳定的.

B.2 稳定性基本定理

在介绍李雅普诺夫稳定性基本定理之前, 先给出函数半正定和正定的定义.

> **定义 B.4 (实函数的半正定性 (positive semi-definite))**
> 设 $\boldsymbol{x} \in \mathbf{R}^n$, Ω 是 $\mathbf{R}^n$ 中包含原点的一个区域, 若实函数 $V(\boldsymbol{x})$ 对任意 n 维非零向量 $\boldsymbol{x} \in \Omega$ 都有 $V(\boldsymbol{x}) \geq 0$, 且 $V(\boldsymbol{0}) = 0$, 则称函数 $V(\boldsymbol{x})$ 为区域 Ω 上的半正定函数.

> **定义 B.5 (实函数的正定性 (positive definite))**
> 设 $\boldsymbol{x} \in \mathbf{R}^n$, Ω 是 $\mathbf{R}^n$ 中包含原点的一个区域, 若实函数 $V(\boldsymbol{x})$ 对任意 n 维非零向量 $\boldsymbol{x} \in \Omega$ 都有 $V(\boldsymbol{x}) > 0$; 当且仅当 $\boldsymbol{x} = \boldsymbol{0}$ 时, 才有 $V(\boldsymbol{x}) = 0$, 则称函数 $V(\boldsymbol{x})$ 为区域 Ω 上的正定函数.

例如, 在 $\mathbf{R}^2$ 中, $2x_2^2$ 和 $(x_1 + x_2)^2$ 都是半正定函数, 而 $x_1^2 + x_2^2$ 和 $(x_1 - x_2)^2 + x_2^2$ 都是正定函数.

注 我们说实函数 $V(\boldsymbol{x})$ 是半负定的, 如果 $-V(\boldsymbol{x})$ 是半正定的; $V(\boldsymbol{x})$ 是负定的, 如果 $-V(\boldsymbol{x})$ 是正定的.

> **定理 B.1 (稳定性定理)**
> 设微分系统(B.1)的平衡点 $\boldsymbol{x}_e = \boldsymbol{0}$, 若存在一个有连续一阶偏导数的正定函数 $V(\boldsymbol{x})$, 使得 $\dot{V}(\boldsymbol{x})$ 为半负定的, 则该系统在原点处的平衡点是稳定的.

> **定理 B.2 (渐近稳定性定理)**
> 设微分系统(B.1)的平衡点 $\boldsymbol{x}_e = \boldsymbol{0}$, 若存在一个有连续一阶偏导数的正定函数 $V(\boldsymbol{x})$, 使得 $\dot{V}(\boldsymbol{x})$ 为负定的, 则该系统在原点处的平衡点是渐近稳定的. 更进一步地, 若随着 $\|\boldsymbol{x}\| \to \infty$, 有 $V(\boldsymbol{x}) \to \infty$, 那么该系统在原点处的平衡点是大范围(全局)渐近稳定的.

李雅普诺夫稳定性定理是判别系统稳定性的充分条件, 而非必要条件. 也就是说, 若找到满足一定条件的一个李雅普诺夫函数, 则可判断平衡点是稳定的或渐近稳定的. 但是, 如果一时找不到这样的李雅普诺夫函数, 也并不意味着平衡点就不是稳定的.

例 B.1 试确定如下状态方程描述的系统的平衡点稳定性:

$$\begin{cases} \dot{x}_1 = x_2 - x_1(x_1^2 + x_2^2) \\ \dot{x}_2 = -x_1 - x_2(x_1^2 + x_2^2) \end{cases}$$

解 显然, 原点 $(0,0)$ 是给定系统的唯一平衡点, 如果选择正定函数 $V(\boldsymbol{x}) = x_1^2 + x_2^2$

为李雅普诺夫函数, 那么沿任意轨迹 $\boldsymbol{x}(t)$, $V(\boldsymbol{x})$ 对时间的导数

$$\dot{V}(\boldsymbol{x}) = 2x_1\dot{x}_1 + 2x_2\dot{x}_2 = -2(x_1^2 + x_2^2)^2 < 0$$

是负定的. 此外, 当 $\|\boldsymbol{x}\| \to \infty$ 时, 有 $V(\boldsymbol{x}) \to \infty$. 因此, 在原点处的平衡点是全局渐近稳定的.

例 B.2 试确定如下状态方程描述的系统的平衡点稳定性:

$$\begin{cases} \dot{x}_1 = x_2 \\ \dot{x}_2 = -x_1 - x_2 \end{cases}$$

解 显然, 原点 $(0,0)$ 是给定系统的唯一平衡点. 如果选择李雅普诺夫函数为 $V(\boldsymbol{x}) = x_1^2 + x_2^2$, 则

$$\dot{V}(\boldsymbol{x}) = 2x_1\dot{x}_1 + 2x_2\dot{x}_2 = -2x_2^2 \leqslant 0$$

是半负定的. 因此, 可以判断此系统在原点是稳定的, 但不能判断其是否渐近稳定. 然而, 如果选择李雅普诺夫函数为 $V(\boldsymbol{x}) = [(x_1+x_2)^2 + 2x_1^2 + x_2^2]/2$, 则

$$\dot{V}(\boldsymbol{x}) = -x_1^2 - x_2^2 < 0$$

是负定的. 因此, 该系统在原点是渐近稳定的.

由以上两个例子可以看出, 用李雅普诺夫方法判断一个系统在平衡点的稳定性, 往往需要选择合适的李雅普诺夫函数.

附录 C 拉萨尔不变原理

本附录主要介绍微分系统定性分析中的另一个重要方法——拉萨尔(LaSalle)不变原理, 其与李雅普诺夫方法相似, 但也有很多不同的地方. 首先从一个具体的例子来进行分析.

例 C.1 考虑一个没有外力的弹簧质量系统, 其状态方程为

$$\begin{cases} \dot{x}_1 = x_2 \\ \dot{x}_2 = -bx_2 - kx_1 \end{cases}$$

其中 $b > 0, k > 0$, 总能量可以表示为

$$V(x_1, x_2) = \frac{1}{2}x_2^2 + \frac{1}{2}kx_1^2$$

进而

$$\dot{V}(x_1, x_2) = x_2\dot{x}_2 + kx_1\dot{x}_1 = -bx_2^2 \leqslant 0$$

是半负定的, 但除 $x_2 = 0$ 外满足 $\dot{V} < 0$. 系统如果要保持 $\dot{V}(\boldsymbol{x}) = 0$, 则系统轨线必须限定在 $x_2 = 0$. 而实际上是不可能的, 除非 $x_1 = 0$. 进而, 由 $x_2(t) = 0$ 可以推出 $\dot{x}_2(t) = 0$, 因此 $bx_2 + kx_1 = 0$, 即 $x_1 = 0$.

如果推导出 $\dot{V}(\boldsymbol{x}) \leqslant 0$, 而且还知道对于 $\dot{V}(\boldsymbol{x}) = 0$ 除原点外, 没有任何系统轨线能永远保留在 $\dot{V}(\boldsymbol{x}) = 0$ 中, 那么就可以判定 $\boldsymbol{x} = \boldsymbol{0}$ 是渐近稳定的. 这种思想就是**拉萨尔不变原理**.

这一部分主要考虑如下的向量形式的自洽微分系统:

$$\frac{\mathrm{d}\boldsymbol{x}}{\mathrm{d}t} = \boldsymbol{f}(\boldsymbol{x}) \tag{C.1}$$

其中 $\boldsymbol{x} \in \mathbf{R}^n$.

在介绍拉萨尔不变原理之前, 先介绍不变集的概念.

定义 C.1 (正极限点 (ω—极限点))

p 称为 $x(t)$ 的正极限点, 如果存在时间序列 $\{t_k\}$, $k \to \infty$ 时, $t_k \to \infty$, 使得 $x(t_k) \to p$.

定义 C.2 (正极限集)

$x(t)$ 的所有正极限点的集合称为 $x(t)$ 的正极限集.

定义 C.3 (不变集)

集合 M 称为微分系统(C.1)的不变集, 如果

$$x(0) \in M \Rightarrow x(t) \in M, \quad \forall t \in \mathbf{R}$$

即: 如果一个解在某一时刻属于 M, 则在所有过去和未来时间均属于 M.

定义 C.4 (正不变集)

集合 M 称为微分系统(C.1)的正不变集, 如果

$$x(0) \in M \Rightarrow x(t) \in M, \quad \forall t \geqslant 0$$

定义 C.5 (集合收敛)

称当 $t \to \infty$, $x(t)$ 趋向于集合 M ($\lim_{t \to \infty} x(t) \to M$), 是指若 $\exists \hat{x} \in M$ 和 $t_k \to \infty$ $(k \to \infty)$ 使得

$$\|x(t_k) - \hat{x}\| \to 0$$

或者

$$\text{dist}(x(t), M) \to 0$$

这里 $\text{dist}(x, M)$ 表示点 x 到集合 M 的距离, 即 x 到 M 中所有点的最小距离: $\text{dist}(x, M) = \inf_{y \in M} \|x - y\|$.

接下来, 给出拉萨尔不变原理的主要结论.

定理 C.1 (拉萨尔不变原理)

设 Ω 是 $\mathbf{R}^n$ 中的有界闭集, 从 Ω 内出发的微分系统(C.1)的解对于 $t > 0$ 均停留在 Ω 内, 如果存在函数 $V: \Omega \to \mathbf{R}$ 具有连续偏导数, 在 Ω 内 $\dot{V}(x) \leqslant 0$, 又设

$$E = \{x | \dot{V}(x) = 0, x \in \Omega\}$$

$M \subset E$ 是 E 中的最大不变集, 则对于 $\forall x_0 \in \Omega$, 微分系统(C.1)的解轨线满足 $\lim_{t \to \infty} \boldsymbol{x}(t) = M$.

拉萨尔不变原理放松了对于李雅普诺夫定理中 $\dot{V} < 0$ 负定的要求, 而且也没有要求函数 $V(\boldsymbol{x})$ 是正定的. 如果 E 中仅含唯一的平衡点, 则该平衡点是渐近稳定的.

回顾例C.1中的问题, 由于 $\dot{V}(x_1, x_2) = -bx_2^2 \leqslant 0$, 可以得出系统的平衡点 $(0,0)$ 是稳定的. 进而, 当 $\dot{V}(x_1, x_2) = 0$ 时, 可以推出 $x_1 = x_2 = 0$, 即拉萨尔不变原理中的集合 $E = \{(0,0)\}$, 因此可以得出系统的平衡点 $(0,0)$ 也是吸引的.

注 虽然上述的拉萨尔不变原理是针对连续的微分系统给出的结论, 但是对于离散系统

$$\boldsymbol{x}(k+1) = \boldsymbol{f}(\boldsymbol{x}(k))$$

也是适用的. 只要将拉萨尔不变原理中的 $\dot{V}(\boldsymbol{x})$ 替换成 $V(\boldsymbol{x}(k+1)) - V(\boldsymbol{x}(k))$, 便可以得到类似的结论. 更详细的内容读者可以参考文献[5].

附录 D　机器人控制器设计

这一部分介绍实现机器人按照指定点移动的控制器设计方法. 考虑二维平面上移动的两轮机器人, 其只有在 x 轴方向上的线速度和绕 z 轴的旋转角速度, 并且满足运动学方程:

$$\begin{cases} \dot{x} = v\cos\theta \\ \dot{y} = v\sin\theta \\ \dot{\theta} = \omega \end{cases}$$

其中 v 表示机器人前行的线速度, ω 表示机器人转动的角速度, θ 表示机器人的姿态, 即机器人的朝向.

假设机器人移动的目标位置坐标是 $(\hat{x}, \hat{y})$, 机器人当前的实际位置坐标是 (x, y), 将设计合适的控制器控制机器人移动的线速度 v 和角速度 ω, 促使得机器人能够向目标位置移动. 记机器人当前位置与目标位置的距离为 ρ, 即

$$\rho = \sqrt{(x-\hat{x})^2 + (y-\hat{y})^2}$$

角度差为 η, 即 $\eta = \arctan((y-\hat{y})/(x-\hat{x}))$ (如图D.1所示). 因为机器人当前的姿态是 θ, 因此机器人移动到目标位置的姿态偏差是 $\eta - \theta$, 记为 ζ, 即 $\zeta = \eta - \theta$. 通过简单的推导, 可以得到机器人从当前位置移动到目标位置的微分动态系统:

$$\begin{cases} \dot{\rho} = -v\cos\zeta \\ \dot{\zeta} = \dfrac{v\sin\zeta}{\rho} - \omega \end{cases} \tag{D.1}$$

关于机器人运动控制器的设计有很多种方法, 感兴趣的读者可以阅读文献[13, 15], 这里我们选择文献[15]的控制器:

$$\begin{cases} v = k_v \rho \cos\zeta \\ \omega = k_\omega \zeta + k_v \cos\zeta \sin\zeta \end{cases} \tag{D.2}$$

其中 $k_v > 0$ 和 $k_\omega > 0$ 是线速度和角速度的控制系数. 下面将利用李雅普诺夫稳定性理论证明系统(D.1)在控制器(D.2)作用下其平衡点 $(\rho, \zeta) = (0, 0)$ 是渐近稳定的.

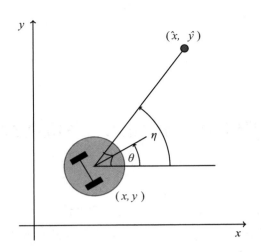

图 D.1   机器人当前位置与目标位置示意图

事实上, 构造如下的李雅普诺夫函数:

$$V(\rho, \zeta) = \frac{1}{2} k_v \rho^2 + \frac{1}{2} k_\omega \zeta^2$$

则

$$\begin{aligned} \dot{V}(\rho, \zeta) &= k_v \rho \dot{\rho} + k_\omega \zeta \dot{\zeta} \\ &= -k_v \rho v \cos \zeta + k_\omega \zeta \left(\frac{v \sin \zeta}{\rho} - \omega \right) \end{aligned}$$

将控制器(D.2)代入上式, 可得

$$\dot{V}(\rho, \zeta) = -k_v^2 \rho^2 \cos^2 \zeta - k_\omega^2 \zeta^2 \leqslant 0$$

从上式可以看出, 当且仅当 $\rho = \zeta = 0$ 时, 有 $\dot{V}(\rho, \zeta) = 0$, 即 $\dot{V}(\rho, \zeta)$ 是负定的. 因此, 根据李雅普诺夫渐近稳定性定理或者拉萨尔不变原理都可以得出微分动态系统(D.1)的平衡点 $(\rho, \zeta) = (0, 0)$ 是渐近稳定的, 即在控制器(D.2)作用下, 机器人可以趋近于目标位置.

参 考 文 献

[1] Wang B, Sun S, Ren W. Distributed Continuous-Time Algorithms for Optimal Resource Allocation with Time-Varying Quadratic Cost Functions[J]. IEEE Transactions on Control of Network Systems, 2020, 7(4): 1974-1984.

[2] Kinderlehrer D, Stampacchia G. An Introduction to Variational Inequalities and Their Applications[M]. New York: Academic, 1982.

[3] Derenick J C, Spletzer J R. Convex Optimization Strategies for Coordinating Large-Scale Robot Formations[J]. IEEE Transactions on Robotics, 2007, 23(6): 1252-1259.

[4] LaSalle J P. The Stability of Dynamical Systems[M]. Philadelphia: SIAM, 1976.

[5] Fazlyab M, Paternain S, Preciado V M, et al. Prediction-correction interior-point method for time-varying convex optimization[J]. IEEE Transactions on Automatic Control, 2018, 63(7): 1973-1986.

[6] Fukushima M. Equivalent Differentiable Optimization Problems and Descent Methods for Asymmetric Variational Inequality Problems[J]. Mathematical Programming, 1992, 53(1): 99-110.

[7] Liu Q, Wang M. A Projection-Based Algorithm for Optimal Formation and Optimal Matching of Multi-Robot System[J]. Nonlinear Dynamics, 2021, 104: 439-450.

[8] Olfati-Saber R, Murray R M. Consensus Problems in Networks of Agents with Switching Topology and Time-Delays[J]. IEEE Transactions on Automatic Control, 2004, 49(9): 1520-1533.

[9] Boyd S, Vandenberghe L. 凸优化 [M]. 王书宁, 许鋆, 黄晓霖, 译. 北京: 清华大学出版社, 2013.

[10] Macenski S, Martín F, White R, et al. The Marathon 2: A Navigation System[C]//2020 IEEE/RSJ International Conference on Intelligent Robots and Systems (IROS). 2020: 2718-2725.

[11] Ren W, Cao Y. Distributed Coordination of Multi-agent Networks: Emergent Problems, Models, and Issues[M]. London: Springer-Verlag, 2011.

[12] Liu W, Wang X, Li S. Formation Control for Leader-Follower Wheeled Mobile Robots Based on Embedded Control Technique[J]. IEEE Transactions on Control Systems Technology, 2023, 31(1): 265-280.

[13] 丁丽娟, 程杞元. 数值计算方法 [M]. 北京: 高等教育出版社, 2011.

[14] 刘彤, 宗群, 刘朋浩, 等. 基于结构持久图和视觉定位的多机器人编队生成与控制 [J]. 信息与控制, 2018, 47(3): 314-323.

[15] 孙文瑜, 徐成贤, 朱德通. 最优化方法 [M]. 2 版. 北京: 高等教育出版社, 2010.

[16] 廖晓昕. 稳定性的理论、方法和应用 [M]. 2 版. 武汉: 华中科技大学出版社, 2010.

[17] 陈宝林. 最优化理论与算法 [M]. 2 版. 北京: 清华大学出版社, 2005.

索 引

Slater 条件, 7

不变集, 221
不稳定, 217
严格凸函数, 5
严格单调映射, 8

元素, 2
全局渐近稳定, 217
凸函数, 5
凸集, 2
动作 (action) 通信, 61
动力系统, 212
半正定函数, 218
单调变分不等式, 8
单调映射, 8
变分不等式, 8
可行域, 7
吸引, 217
定位, 172
对偶问题, 10
平衡态, 216
平衡点, 216
形状图标, 26
微分动力系统, 212
微分方程, 213
微分方程组, 213

投影算子, 3
拉格朗日函数, 9
拉萨尔不变原理, 220
收敛, 217
时变优化, 37
最速下降法, 6
服务 (service) 通信, 56
李雅普诺夫第一法, 216
李雅普诺夫第二法, 216
梯度下降法, 6
正不变集, 221
正定函数, 218
正极限点, 221
正极限集, 221
消息 (message) 通信, 49
渐近稳定, 217

稳定, 217

闭凸集, 2
队形, 11
队形保持, 11
队形切换, 11
队形等价类, 26
集合, 2
鞍点, 9
鞍点定理, 10

后　　记

　　多机器人协同是人工智能领域的一个新型研究方向,其实现途径不仅仅局限于优化方法,感兴趣的读者可以进一步阅读这一方向的其他研究方法,并尝试使用本书中的代码结构对算法进行实践.

　　多机器人协同的研究除了本书中提到的领导–跟随、编队和围堵问题,还有协同感知、协同避障、协同路径规划等研究方向,这些也是多机器人协同研究中的重要课题,具有一定的挑战性,其理论和实践研究都需要进一步发展和完善.

　　本书只使用了非线性优化方法实践多机器人系统的协同控制. 希望本书能够起到抛砖引玉的作用,能够帮助读者发掘并实现更多、更先进的多机器人协同控制方法.

编　者

2023 年 5 月